Easi-Bild SIMPLIFIES ELECTRICAL REPAIRS

by Donald R. Brann

Library of Congress Card No. 70-95701

FIFTH PRINTING – 1973

REVISED EDITION

Published by
DIRECTIONS SIMPLIFIED, INC.
Division of
EASI-BILD PATTERN CO., INC.
Briarcliff Manor, N.Y. 10510

FIRST PRINTING
Copyright, 1970
SECOND PRINTING – REVISED EDITION
Copyright, 1971
THIRD PRINTING – REVISED EDITION
Copyright, 1972
FOURTH PRINTING – REVISED EDITION
Copyright, 1973

NOTE

Due to the variance in quality and availability of many materials and products, always follow directions a manufacturer and/or retailer offers. Unless products are used exactly as the manufacturer specifies, its warranty can be voided. While the author mentions certain products by trade name, no endorsement or end use guarantee is implied. In every case the author suggests end uses as specified by the manufacturer prior to publication.

Since manufacturers frequently change ingredients or formula and/or introduce new and improved products, or fail to distribute in certain areas, trade names are mentioned to help the reader zero in on products of comparable quality and end use.

The Publisher.

LEARNING TO LIVE BETTER

We live in a strange and wondrous age of miracles. By harnessing the gravitational power of a planet, men discovered how to travel further, faster, and finally to walk on the moon. Now, another new, and equally important element, the magnetic force that surrounds an inanimate object is being discovered.

A case in point is an appliance, fixture or switch that fails to function. It immediately radiates small waves of irritation that stimulate larger currents of despair. The brain cell, influenced by both negative and positive thoughts, continually seeks to solve even the smallest problem. Each one it solves stimulates positive waves of self-esteem, while those it fails to resolve, creates frustration.

Those who have learned something of the art called living, believe problems contain activated particles that keep us young. By continually facing and solving problems, they generate a magnetic zest for living. Those who don't try, begin to die long before they have learned to live.

Learning to solve electrical problems while all circuits are operative, is relatively easy. It's like learning to swim in calm, shallow water. Figuring out what needs to be done, after a circuit goes dead, is like learning to swim after falling into deep water. Read this book and you'll learn to live electrically, a lot more economically.

Don R. Brann

TABLE OF CONTENTS

USE CAUTION AND COMMON SENSE

While many building codes require house wiring be done by a licensed electrician, there is no law that prevents a homeowner from doing what he wants to do. The directions in this book provide a guide that illustrates how and where a licensed electrician does the work. Knowing how, permits the homeowner to prepare the area, drill holes where needed, do that part of the work he wants.

Homeowners who wish to do their own electrical work must follow code approved procedures, use code accepted materials, and where required, apply for a permit. If the experienced homeowner then wants to do the same work for others, he would require a license.

Don't let anyone con you into thinking your work can't meet code requirements. If you use the same materials, and install them in the same way as existing electrical work, no one will be able to tell who did what or when.

If you are adding one or more outlets, or fixtures, to an existing circuit wired with #14 gauge wire, don't extend the circuit with #12 wire. Always use same gauge as existing wire. Only when installing a new circuit from a fuse panel, can you use #12 or other gauge wire.

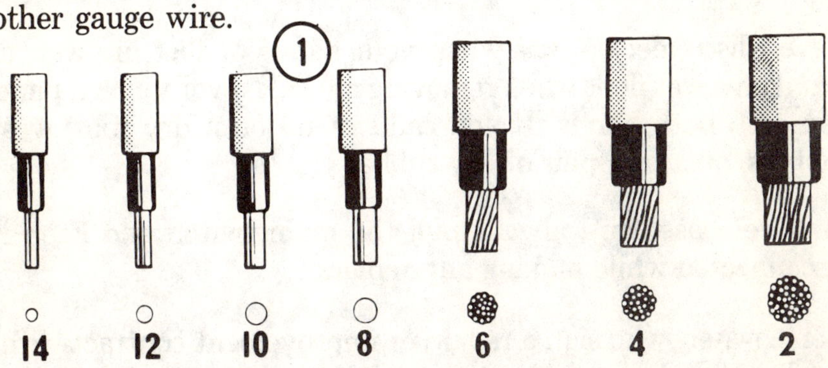

Use copper wire rather than aluminum. Although copper will cost more for the wire itself, it is virtually trouble-free. Aluminum requires special connections and above average skill to insure permanent, trouble-free connections.

Illus. 1. Shows various size wire. While #12 is larger than #14, if you have #14 coming from panel box, it can only transmit, according to codes, 15 amp service. If you didn't know, and used #12 wire to extend service, you still only get 15 amp.

Consider wire as water pipe. If you use ½″ pipe from the main supply line, then extend the line with ¾″, you still can't get more than that supplied by ½″ pipe.

On the other hand, if you are installing a new circuit, and use No. 12 wire from service panel, where other existing circuits are wired with No. 14; or install size wire new equipment requires, and use fuse recommended for that wire, you will make an installation no one can fault. Most codes now require circuit breakers.*

Electrical maintenance and cooking have much in common. You don't put your hand over a hot flame, and you don't hold a hot wire. Always exercise caution. While an experienced electrician can make some repairs with the current connected, don't attempt any repairs until you either pull the main switch, or disconnect the correct fuse. And always make certain you know which fuse is the right one. You can then close the main switch and keep service operative for the balance of the house.

Never disconnect a fuse when your hands or feet are wet, or stand on a wet floor when removing a fuse. Never touch a panel box with both hands. If you can't stand on a dry spot, wear rubbers, and/or a pair of dry rubber gloves.

In a wet basement, always pull the main switch and keep it disconnected while making any repairs.

Homeowners who sign a repair, or improvement contract without knowing how much material and time the work requires, are buying a pig in a poke. If a skilled craftsman says he doesn't know how long a job will take — beware! If he's never done a similar job, he could be telling the truth. In this case, consider whether you want to do some of the preliminary work yourself.

* Modifications required to conform to the Canadian Electrical Code- See pages 97, 98.

To intelligently buy any installation or service, first learn how, where, and what needs to be done. Obtain a price on materials and equipment, separate from labor. Be sure the material and equipment you shop for, meets local code requirements. Be sure materials and equipment an electrician supplies also meet local codes. Be sure the necessary amount of copper, size of wire, is actually installed.*

Knowing how a job is done permits intelligent discussion concerning length of time it will take to do what you want done. If you know, and the electrician knows you know, a meeting of the minds takes place, both parties benefit. When you do all, or only part of the work, considerable savings can be effected.

An installation can be made with a size wire that works, but actually doesn't provide all the power the equipment requires. Years ago, I needed a 230 volt outlet approximately 100 feet from the fuse panel. Quotations received from two well-established electrical contractors were close. The installation was made and the equipment put into service during my absence.

While it was operative, it continually gave trouble. Months later, when I carefully examined the installation, I realized the wire size was inadequate for distance involved. The loss of power was the cause of the trouble.

In new construction an electrical contractor will usually count the number of outlets the job requires, multiply this by a fixed cost per outlet, and quote total cost. In this case, they figure a switch as an outlet. If you question total cost, measure length of cable required, count the number of boxes, outlets, and switches the job requires. Get a price on all materials, then subtract it from estimate submitted.

If you are modernizing a basement, attic, or new addition, plan one circuit for every 400 square feet. Install an appliance circuit with #12 wire; this permits a 20 amp fuse which provides 2300 or 2400 watts. Install lighting circuits with #14 wire. Use 15 amp fuse with #14 wire.* Do not use appliance circuit for lighting.

* Modifications required to conform to the Canadian Electrical Code-
See pages 97, 98.

Install wall outlets for every 6′ of wall space; one for every 3′ of counter space in a kitchen. Note where they are placed in a room that's comfortable to use. Never install a wall outlet where it can be reached from a bathtub or shower.

Electrical repairs, like changing a spark plug, follows established procedures. Anything that needs to be replaced, simply reverses the procedure of installation. Everything one man can put together, another can usually take apart, then reassemble. But play safe, don't trust your memory. Always make a check list, or where there are many parts, place them on a large calendar, or in a numbered muffin cooking tray. Place the first part you take off on 1, the second part on 2, and so on. When you reassemble, start with the highest number and work back to 1. While the fear of making a mistake frequently throttles normal thinking, numbering each part frees your mind.

LEARN, LOCATE AND LIST

Every house is connected to the utility line with a service entry, Illus. 2. This is either a two or three wire overhead, or underground cable, that runs from the utility pole to the house. Rubber covered wires, (conductors), carry the service through a pipe (conduit), to the meter. Meters are usually located on outside of new houses where utility company specifies. Feeders from the meter carry the service to the main distribution center, the service panel.

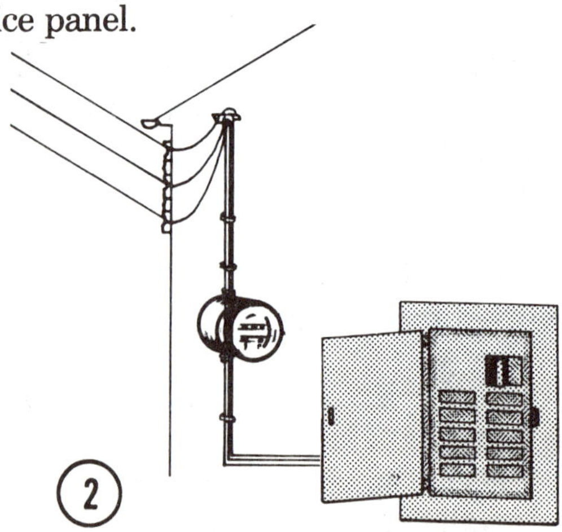

2

To intelligently make electrical repairs that can, through the years, save hundreds of dollars, become familiar with the service panel. Since each room is served by two circuits, knowing which fuse needs replacement is important. Some fixtures may be connected to a circuit serving a different floor.

The easiest time to learn which is what, is when the circuit is operative. Take time to plug lamps into every outlet in a room. Switch on all switches. Be sure to do the same in closets. When checking a bathroom, don't forget to put a night light in the outlet in the medicine chest. With all switches on, and lamps plugged into every outlet, and all burning bright, remove one fuse serving that room.

Make a list noting which outlets and switches are controlled by fuse #1. After making this list, replace fuse #1, switch off all lights and follow the same procedure to locate and to list the outlets and fixtures controlled by fuse #2. If you know which fuse controls each circuit, and what's on a circuit, finding a short in that circuit is greatly simplified. For more information on finding a short circuit, see page 19.

If a fixture being replaced isn't exactly like the replacement, if it has more, or fewer connections, or if the connections are in a new position, ask the retailer to explain how it can be substituted. If you take the faulty fixture, outlet or switch with you, the retailer can explain installation.

Small appliance circuits, those serving the kitchen, laundry, basement workshop, the circuit that serves a refrigerator, toaster, grill, or where similar appliances are likely to be plugged in, should be fused with a 20 amp fuse, and should be wired with a #12 wire. No wire smaller than #12 is acceptable for an appliance circuit.*

If you consider purchasing an older house, and want to make certain house wiring is adequate for the kind of living you have grown accustomed to, ask utility company to tell you how much amperage and voltage is presently available. Note size of fuses currently being used. To make certain the owner is

* Modifications required to conform to the Canadian Electrical Code - See pages 97, 98.

not overloading a circuit, ask an electrician to spot check size of wiring when he finds a 20 amp or larger fuse in the panel box. If in doubt and you can't get help needed from a licensed electrician, ask the present owner to state in writing, size of service, number of 15 amp, 20 amp and heavier circuits the house con-contains.

Rewiring an old house, installing new outlets, or wiring of any kind, can be done easily if you plan on installing new wall paneling and/or ceilings. Just punch holes in existing plaster or wallboard, recess new cable in a slot cut thru plaster, and fasten boxes in position following directions on page 30. If you don't plan on new paneling, the easiest way to install new outlets is by removing the baseboard. Refer to page 42.

Those buying a house with 60 or 100 amp service, must not only increase power to the house, but will need to increase size of distribution lines (circuits) within.

Illus. 3 shows how various switches are connected. L indicates a light fixture. M indicates a motor. B indicates a black (live) wire. W indicates neutral or wire.

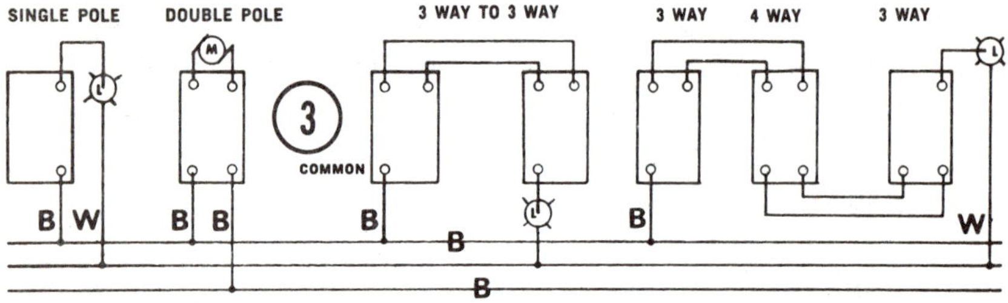

Three way switches permit controlling a light, or series of lights, from two different locations. A four way switch is used if more than two switches are required. For additional information concerning three way switches, refer to page 37.

A double pole switch is connected to two 120 volt lines. Most heavy duty motors require 240 volt service.

240 volt appliances must be wired with size of cable it requires. If in doubt, ask your utility company. Be sure to explain how far from fuse panel the appliance will be placed.

DON'T BE CONNED

If a fuse keeps blowing, don't panic into thinking it's a big deal. Most shorts can be traced and fixed easily if you have mapped your circuits and know which outlet, fixture and switch is on each circuit as suggested on page 19.

Remove blown fuse. DISCONNECT MAIN SWITCH. Using a screwdriver, turn screw in bottom of fuse holder to make certain it's tight. Next screw in a 60 watt lamp in fuse holder. Disconnect all lamp cords, radios, clocks, every plug-in that's on this circuit. Switch off all switches. If there are any three-way switches on this circuit, or pull chain fixtures, remove bulbs. The bulb in the fuse socket should now be out. If it's on, you have a short in that circuit, outlet, fixture or switch.

If the bulb in fuse socket is off, start at the nearest wall outlet and/or switch, or fixture, and try one at a time. If you plug in an extension cord connected to any appliance, and the fuse bulb goes on while the appliance is off, the short could be in the appliance. The bulb should burn bright when you switch the appliance on.

If anyone has recently done any carpentry, the short may be in a nail driven into a cable. If the fuse blew after an electrical storm, it could be moisture or a fused receptacle.

A lamp bulb can cause a short circuit if a piece of filament breaks off and lodges against the other filament.

The bulb in fuse socket should go on every time you switch on a fixture or appliance. It should go out after you switch that fixture off. If it stays on, that switch, outlet, plug or extension cord could be the culprit.

4

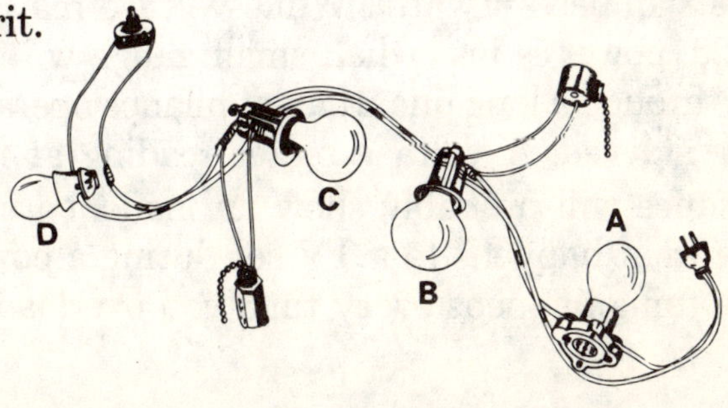

If you begin checking each outlet, switch, fixture, etc., close to the fuse panel, and work out, you can zero in on that part of the circuit where the short may be, if it isn't in a fixture, switch, outlet or bulb.

Illus. 4 helps explain what happens when you place a bulb in a fuse socket to test a circuit. Assume A is the fuse socket. Place 60 watt bulbs in A, B, C, a 10 watt bulb in D. With all switches off, if there's a short in the circuit, Illus. 4 a, bulb A burns full bright. If there's no short, when you switch on B, A and B burn half bright. Switch on C, B and C both get dimmer, A burns brighter. This shows circuit and switches at B and C are operative. Switch on B, C, D and A will burn the brightest while B, C, D are dim. Switch off B and C and A also goes out while D burns bright. The 10 watt bulb in D doesn't draw sufficient power to illuminate A. If you have a short in any one switch or fixture it will immediately draw sufficient power to make A burn real bright. By turning off all other switches, disconnecting all cords, you zero in on the exact trouble spot.

4a

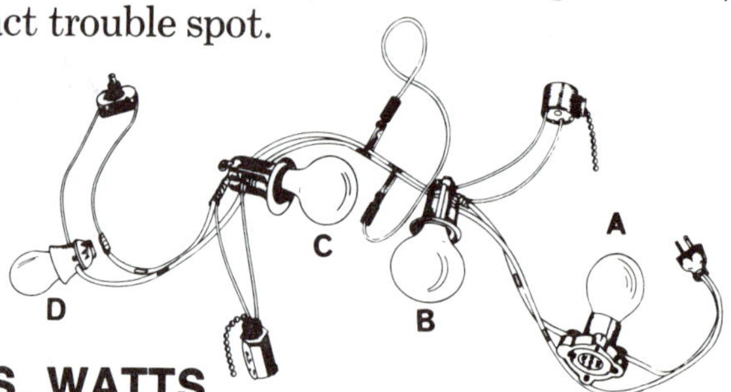

AMPERES, VOLTS, WATTS

Amperes provide a measure of electricity. While it's comparable to gallons when measuring the flow of water, it is not a definite measure, like gallons per minute. Amperes only refer to a flow of electricity, and not an exact amount.

Volts × amperes = watts. While voltage may be rated 110, 115 or 120, power is lost when small gauge wire is used, or when sent through a long line. If an appliance operates sluggishly, it's being starved. A voltage meter reading at receptacle serving appliance will probably show an inadequate source of power. Note what happens to a TV set during a power shortage. The same thing happens every time you overload a circuit.

The National Electrical Code specifies that light circuits be wired with No. 14 wire (minimum size acceptable), and be fused with a 15 amp fuse. If you are building a new addition, or transforming a basement, attic, or garage into living space, never use any size smaller than No. 14 wire for light circuits, No. 12 wire when installing an appliance circuit. No. 12 wire permits using a 20 amp fuse. 20 X 120 volts = 2400 watts. Many codes now specify #12 wire as a minimum for branch circuits. If the run is 50 ft. or more, use #12 wire.*

Line voltage was formerly rated 110 volts. Today most companies provide 115/130 volts. In many old houses, two wire service only provided 30 amps, Illus. 5*. During the late thirties and forties builders began to install 60 amp, 3 wire service. In the fifties, 100 amp, three wire service was usually specified. Today, a minimum of 150 to 200 amps is required to power the appliances most people want.

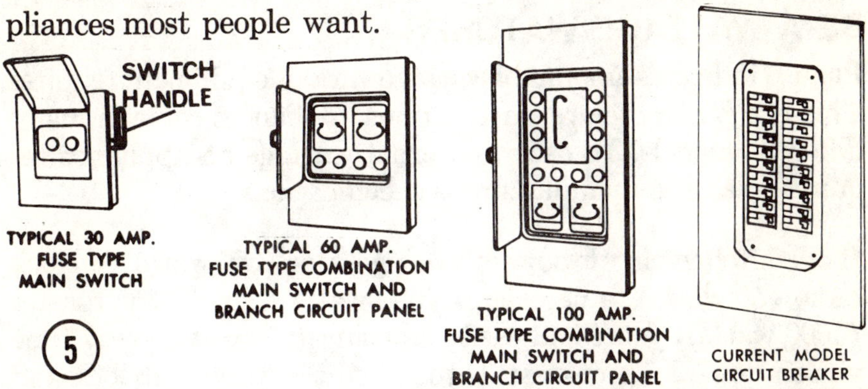

5

240 volt appliances must be wired with size of cable it requires. If in doubt, ask your utility company. Be sure to explain how far from fuse panel the appliance will be placed.

Before buying a new air conditioner, hot water heater, clothes dryer, or other large appliance, and before making a decision concerning a major remodeling job, ascertain existing house power. Estimate amount of wattage currently being used, and how much additional power the proposed equipment requires.

A two wire circuit contains a black and white wire and provides 120 volts. A three wire circuit contains one black and one red

* Modifications required to conform to the Canadian Electrical Code - See pages 97, 98.

120 volt lines, and one white neutral (wire). The two hot lines are split at the panel box to provide two separate 120 volt circuits. When 240 volts are needed for a range, hot water heater, etc., the two hot lines are connected to the appliance, plus a third wire. Range cords, Illus. 86, usually have two brass and one silver colored connectors.

3 wire, 100 amp service normally provides (100 × 230 or 100 × 240) 23,000 or 24,000 watts. A 150 amp, 3 wire service, offers up to 36,000 watts. Those building in an area where electric heat is within their budget, should consider installing 200 amp service. Note: Power to your house may vary from 208 to 240 volts. Always use the lower figure in estimating how much wattage a circuit can supply. While service delivers a normal overload, during peak periods you may not obtain even the lower figure.

BE A WATTAGE WATCHER

Page 97 indicates amount of wattage various equipment requires. This is a general figure since various appliances come in many different models. To ascertain exact wattage an appliance requires, read the manufacturer's specifications.

If one kitchen circuit powers a refrigerator (300 watts), a clock radio (50 watts), a percolator (600 watts), an electric toaster (1000 watts), you could easily be using 1950 watts every time you sit down for breakfast. If the circuit is wired with #12, and fused with a 20 amp fuse, or circuit breaker, it could supply 2300 to 2400 watts. This would be O.K., and all appliances should perform as you and the manufacturer expect. But supposing somebody plugs in an electric iron to hurriedly get a dress ready for school. The average iron draws 1000 watts. This would blow a fuse or break a circuit breaker.*

A #15 amp fused #14 wire circuit provides 1350 watts for approximately 30 ft. with only a 2% drop in voltage; #12 wire fused with a 20 amp fuse, approximately 2000 watts; a 30 amp fuse, #10 wire circuit provides up to 3600 watts for the same distance. A #8 wire will provide comparable wattage up to 60 feet.

* Modifications required to conform to the Canadian Electrical Code - See pages 97, 98.

Illus. 1 shows the actual size of copper wire with the insulation removed.

Installing a 20 amp or 30 amp fuse in a circuit wired with #14 gauge, is like filling a burning fireplace chock full of dry kindling, removing the fire screen and not expecting trouble. When you apply more amperage than the wire was designed to carry, you apply excessive heat. Ascertain what size fuse each circuit requires, and never overload a circuit.

While most appliances are seldom used simultaneously, and most properly wired circuits have a built-in overload, to get the best performance out of each appliance, and/or circuit, use it intelligently.

Each circuit from the panel box requires copper, size of wire, capable of carrying the wattage the distance appliances, fixtures or equipment require. General purpose circuits, those serving lamps, electric shaver, overhead lighting, television, etc., must be fused with 15 amp fuses. A kitchen fan must be installed in a light circuit.

FUSES AND CIRCUIT BREAKERS

Illus. 2, 4, shows the type of fuse panel installed in many homes.

This panel uses the screw-in fuse, Illus. 6. Fuses are designed to prevent overloading a circuit. Like water pressure that bursts a garden hose, heat generated by current passing through an inadequate size wire, can soften and damage insulation. This could become grounded, cause trouble, even start a fire. When you see a wire with damaged insulation, you usually discover the source, or potential source of a short.

6

Fuses are rated according to their capacity. The code requires a 15 amp fuse with #14 wire. A 20 amp fuse with #12 wire. The screw-in fuse, Illus. 6, is the most common found in older houses. Unscrew fuse to replace with a new one.

Another type of screw-in fuse is shown in Illus. 7. This fuse and the fuse adapter, Illus. 8, were designed to prevent overloading a circuit. If anyone in the family, or in a rental unit, continually blow fuses, screw in a 15 amp adapter and fuse in a #14 wire circuit. Be sure to screw adapter all the way in. Once installed, the adapter can't be removed. Since each adapter will only accommodate one size fuse, it eliminates overloading.*

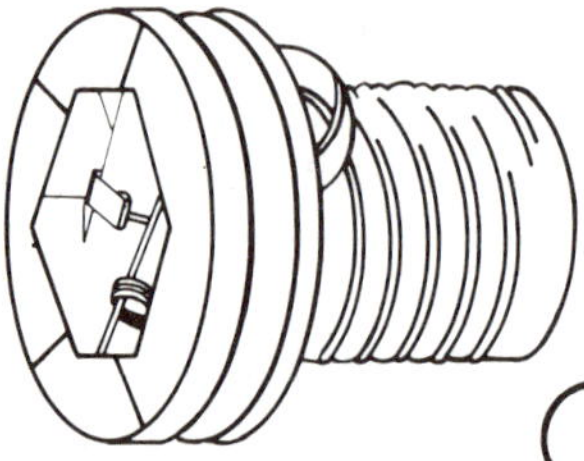

7

8

If you are making a new installation, the local code will specify using this type of fuse and adapter, or a circuit breaker.

Illus. 9 shows the circuit breaker type of panel. When a circuit is overloaded, the breaker trips a switch. To reset, press lever all the way to OFF, then to ON. If line is shorted, it will snap to Off. In this case, follow directions on page 19.

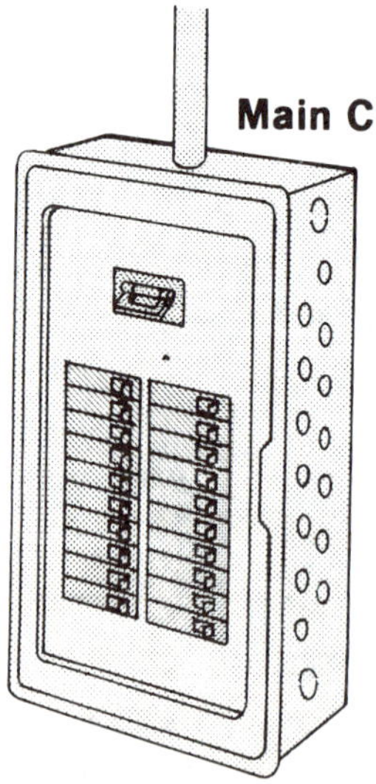

9

* Modifications required to conform to the Canadian Electrical Code - See pages 97, 98.

The cartridge fuse, Illus. 10, is available in various sizes from 10, 15, 20, 25, 30, 35, 40, 45, 50 to 60 amp. These are used to fuse all 240 volt circuits serving a range, hot water heater, and other appliances that require up to 60 amps. Always use exact size fuse equipment requires.

10

Each fuse block, Illus. 11, accommodates two cartridge fuses, Illus. 12, one for each 120 volt circuit. Before removing a fuse block, note how it is installed. Homeowners frequently replace the fuse block upside down. While it fits, it doesn't make contact.*

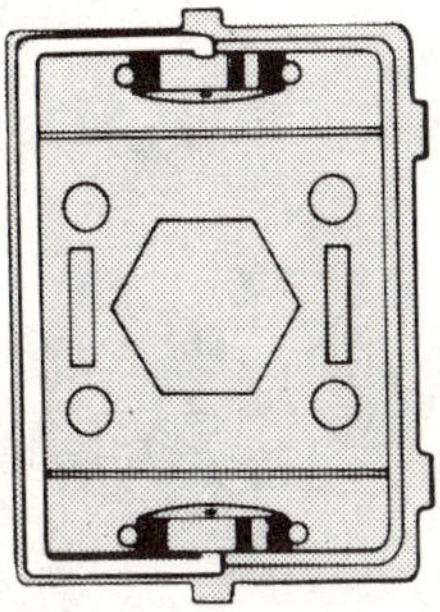

11

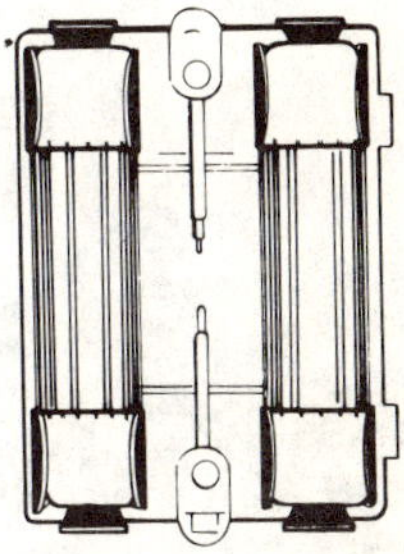

12

If a new tenant in rental property can't get the range or hot water heater to work, inspect the fuse block. Disgruntled tenants frequently reverse position of the fuse block or substitute smaller cartridge fuses. While an electrician can position fuse block properly, and do it in minutes, it's amazing how long it can take to find out what's wrong when someone is paying an hourly rate.

ELECTRICAL PROBLEM #1

Fuses blow when a circuit is overloaded, or when one lamp, or appliance cord, fixture, or appliance is defective. A washing machine that needs lubrication, a motor that needs oil, a toaster that's overloaded with crumbs, can blow a fuse.

* Modifications required to conform to the Canadian Electrical Code - See pages 97, 98.

The first problem that most frequently occurs in every home requires replacing a light bulb or fuse. If a switch, light fixture, or wall outlet fails to function, check the fuse panel. Electricians customarily install outlets in each room on two different circuits. If you blow a fuse on one circuit you still have power on the other circuit. Always make certain you know which fuse controls receptacle you are working on.

If an outlet fails to function, check to see if the outlet is operated by a wall switch. Electricians get frequent calls from homeowners who forget certain outlets are connected to wall switches. Families moving into new homes, or using a familiar room for a new end use, frequently make this mistake.

Outlets have brass terminals, Illus. 13, 14, for the (black) live lead (hot); silver or chrome screws for the white, or neutral wire.

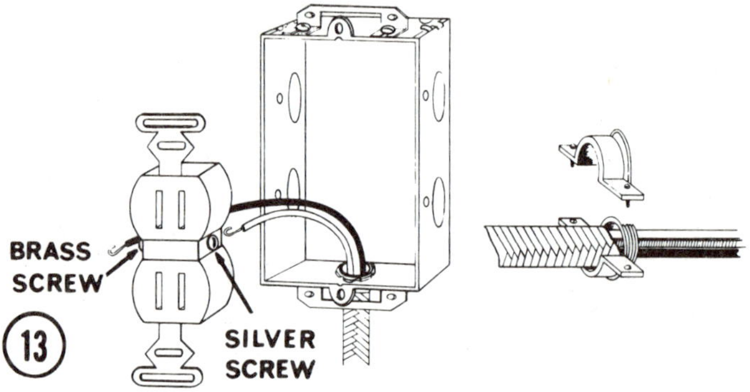

13

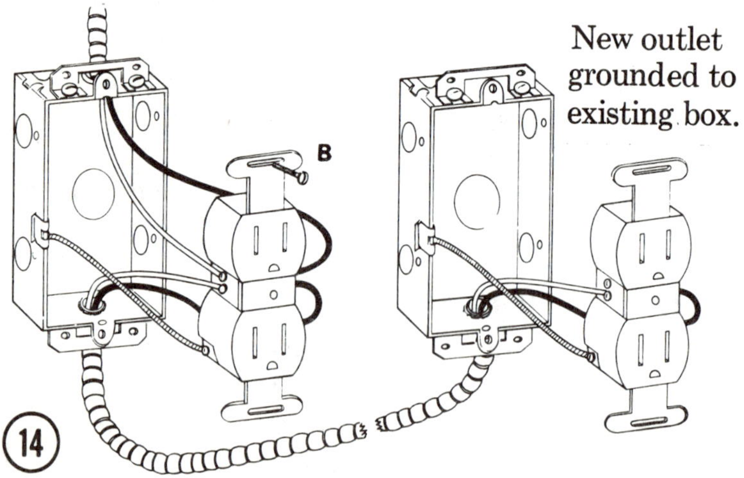

14

A grounded outlet, Illus. 15, has a third terminal. This is either colored green or has a green hexagonal headed screw. In new construction this terminal is grounded to the box with a screw.

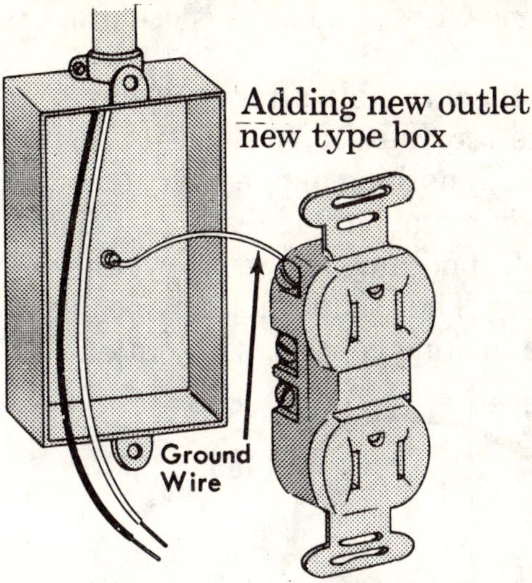

15

It can be grounded with a press-on clip when this type of outlet is installed as a replacement, Illus. 16.* Always use same size wire to ground as was used to other terminals.

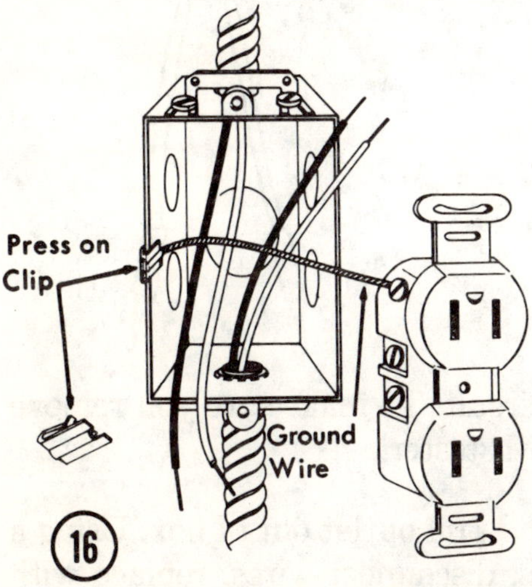

16

17

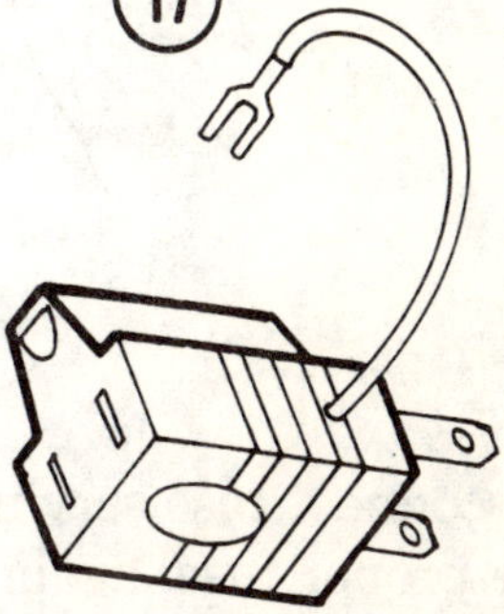

If you want to connect a 3-pronged plug to an existing two slot outlet, use an adapter, Illus. 17.* Be sure to secure ground leg on adapter to screw holding cover plate, and check to be sure screw is of metal.

* Modifications required to conform to the Canadian Electrical Code - See pages 97, 98.

A good fuse and a faulty one are easy to identify by a clouded plastic cover, or broken metal strip; but sometimes, one that doesn't have a blackened window may still be faulty. So always test with a new fuse, and always use proper size.

If you still have trouble after replacing a fuse, insert wires of light bulb tester, Illus. 18, in slots. If it lights, you have current to the outlet. If nothing happens, try one wire in one slot, touch other wire on metal box. If tester now lights you have current to the outlet. If nothing happens, try other slot and touch metal with other leg. The ground slot may be defective, or the ground may not be making contact with outlet.

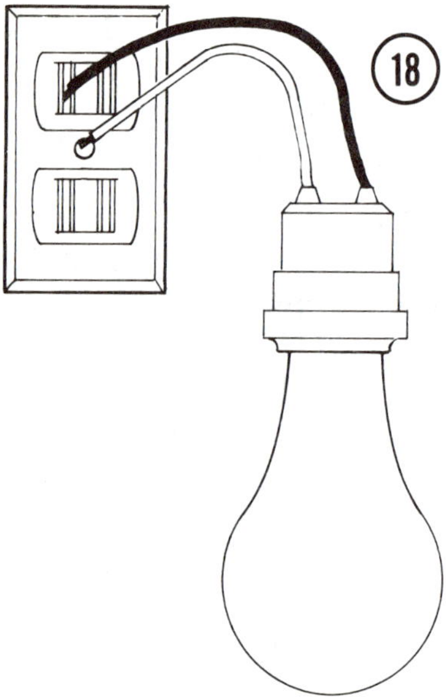

18

19

Disconnect fuse serving this circuit. To make sure you remove the proper fuse, again test with tester.

Remove screws B, Illus. 14, and pull outlet out of box. Using a screw driver, loosen terminals, disconnect wires, replace with an identical outlet.

Another type of tester that's very helpful is shown in Illus. 19. This works the same way as the bulb tester. Always hold insulators when making a test.

CAUTION: Make repairs as an electrician does it. Always strip end of wire clean and bend end into an eye. Place "eye" in position shown. Terminal tightens when turned in direction of curve, Illus. 20.

20

21

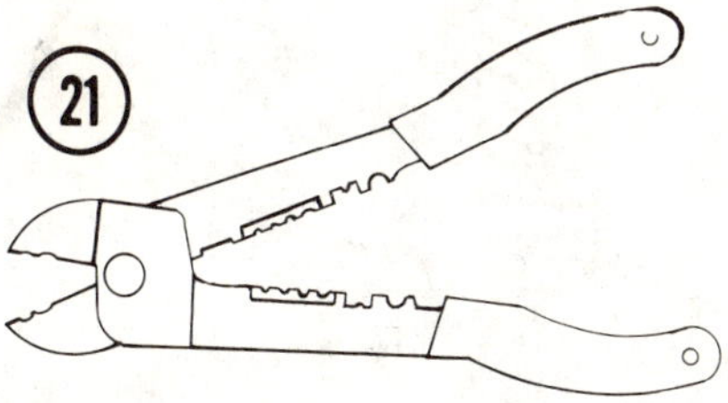

The wire stripping pliers, Illus. 21, simplifies stripping insulation without cutting the wire.

Wall outlets, particularly those frequently used, will in time need replacing. If a lamp begins to flicker, or an appliance doesn't operate properly, try another outlet. If it still acts up, it's the lamp. If it works well in another outlet, the outlet needs replacing. An outlet that serves a vacuum cleaner or other appliance that's constantly being connected and disconnected frequently needs replacing.

A faulty switch can be checked in the following way. Install a new fuse. Test with light tester, Illus. 22. If circuit is operative tester will light when pigtail touches A and the box; or on B and the box when switch is closed. If bulb lights when switch is closed, switch is operative. Try switch several times. A loose contact in switch may only work intermittently.

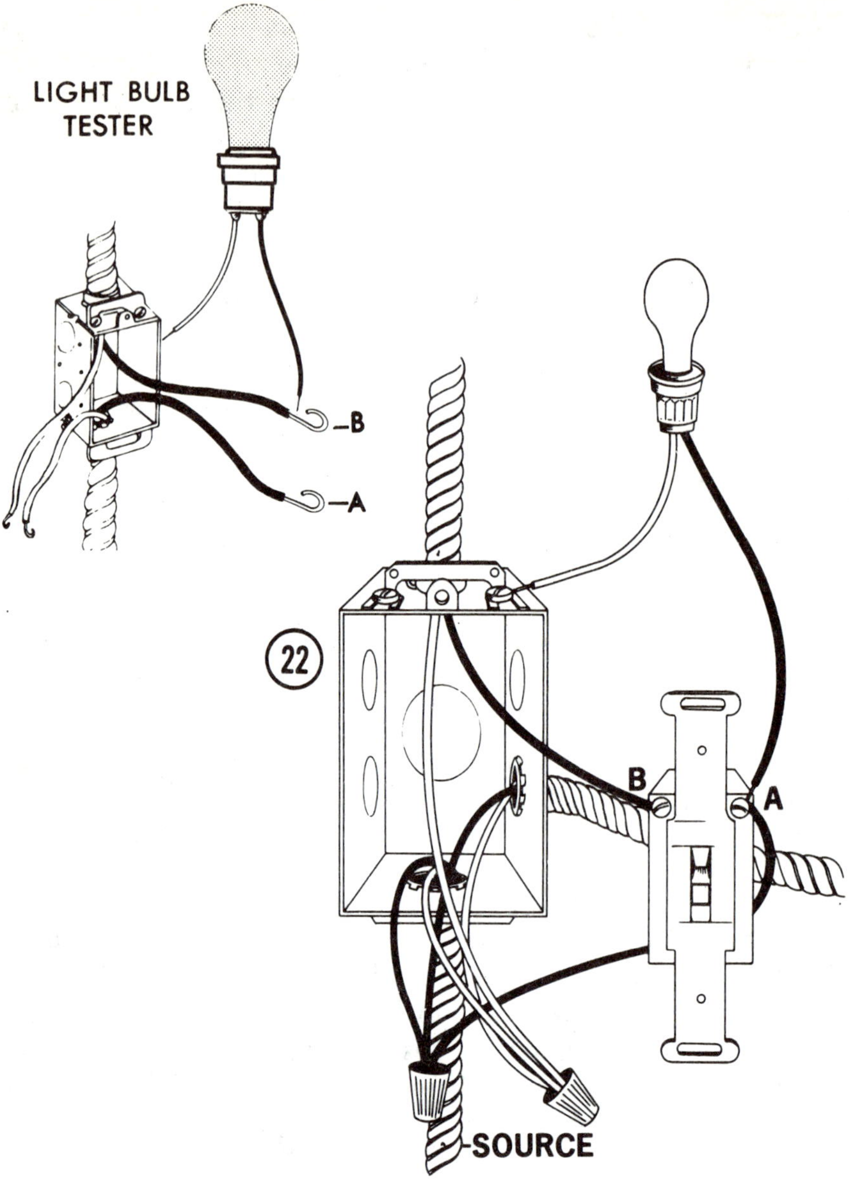

To find out which lead is from source, touch one leg to A, the other leg to box. If bulb lights, A is from source. If it doesn't light, try B and box.

Bulb should light when pigtails touch A and B, Illus. 23. This indicates power to the switch.

CEILING LIGHT REPAIR AND INSTALLATION

A faulty fixture can be checked in the same way. First, test current to switch, then test to see if switch is operative. If you have current, and switch is working, the trouble is usually in the fixture. Remove fuse. Remove globe and fixture from box. Lower fixture, Illus. 24. Replace fuse. Turn switch ON. Test terminals on fixture with bulb tester. If you have power to this point and the fixture still doesn't work, the trouble is in the fixture.

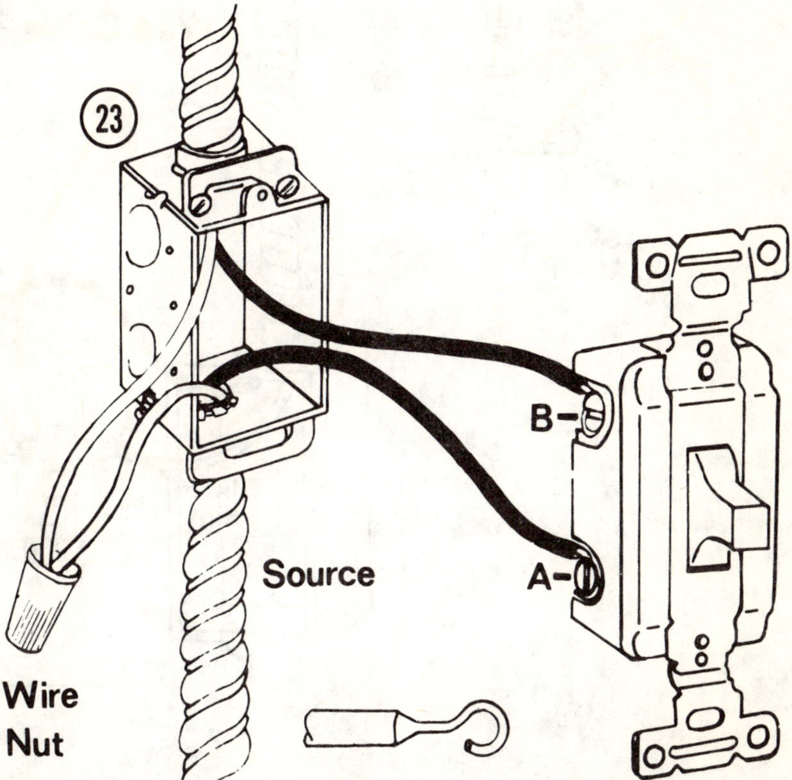

Illus. 24 shows a typical installation of what is called a switch loop found in many bathrooms. Because two-wire cable contains a black and white wire, and a switch is only installed in a live line, both wires in a switch loop are considered live, (black). Black from source enters switch at A and leaves through B. When you use a tester on a switch, you will only find one lead live when switch is off; both terminals live when switch is on.

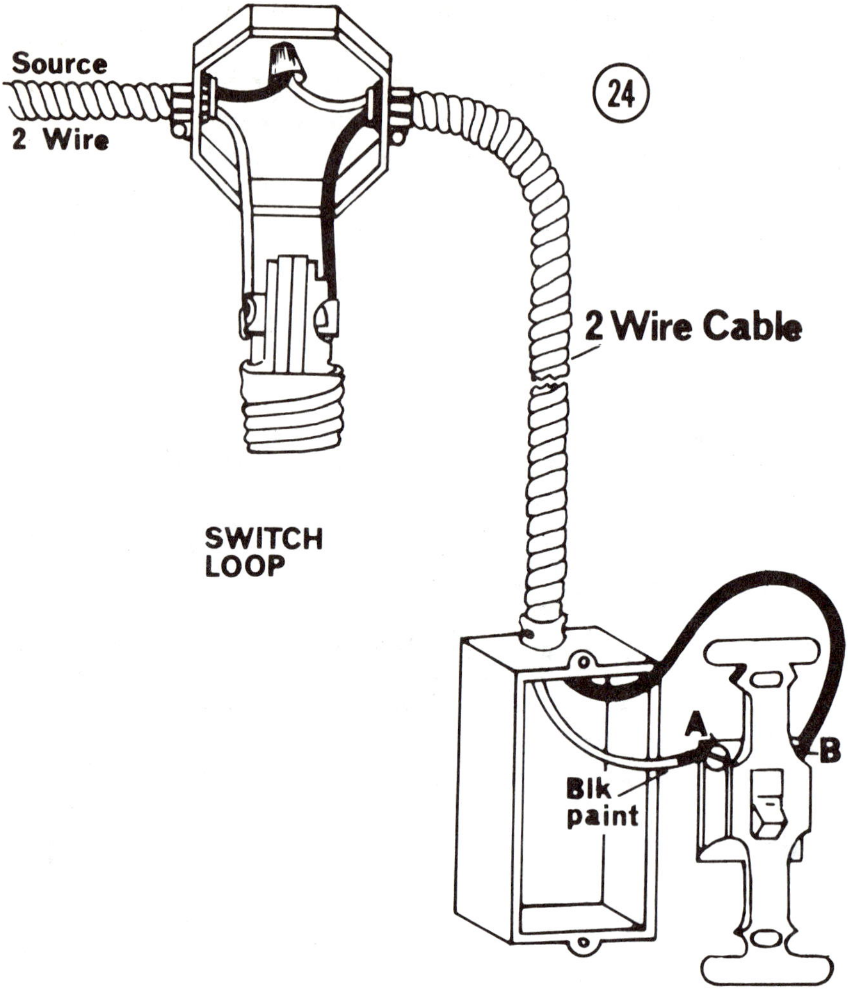

Unscrew ring A, Illus. 25, if it has one. Loosen screws, or nuts at B. Remove base C. Loosen wires at terminals. Replace interior D with identical part. A fixture stud, E, is fastened to box with two screws. Many overhead fixtures are fastened directly to this stud with a threaded stem.

Use a pull chain interior, or a pull chain interior with side outlet where required. Note: Page 98.

25

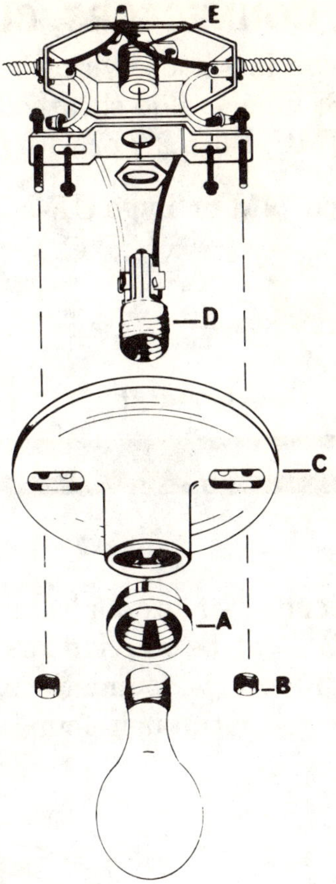

26

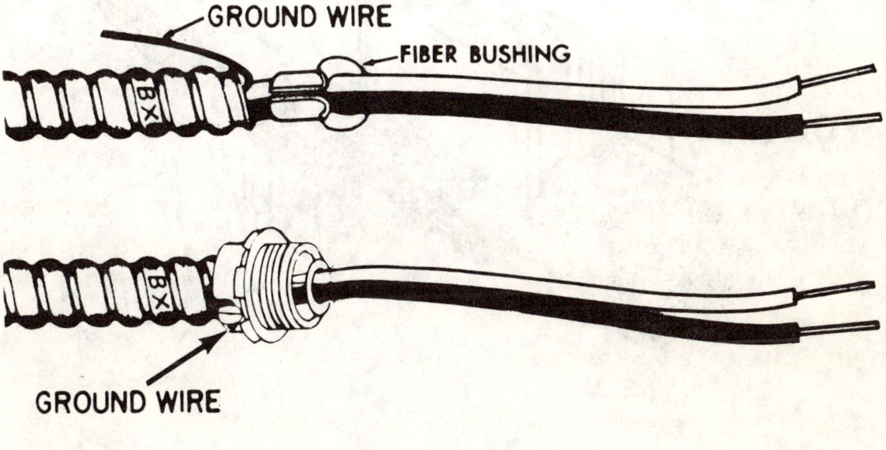

BOXES, CABLE, CONNECTORS, CLAMPS

The National Code specifies house wiring be done with armored cable, BX, Illus. 26; nonmetallic sheathed cable, either Type NM or Type UF-NMC, Illus. 27; or with conduit, Illus. 28.*

Type NM or Type UF-NMC

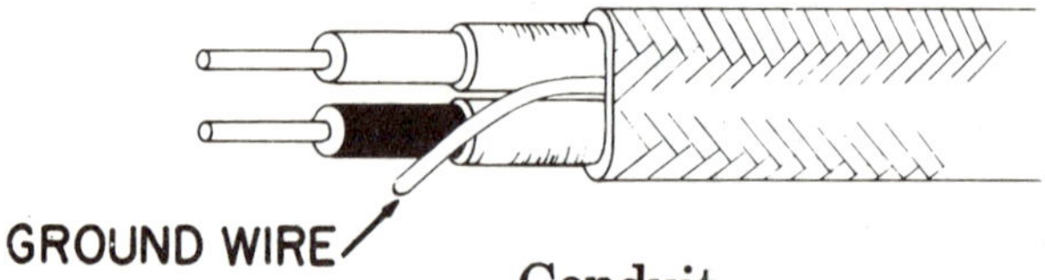

27

Conduit

28

Type NM is recommended for interior use and only in dry areas, while Type UF - NMC can be used indoors as well as out, even in wet areas, through footings, foundations, etc., Illus. 29. The code specifies cable going through foundation be sleeved in conduit.*

29

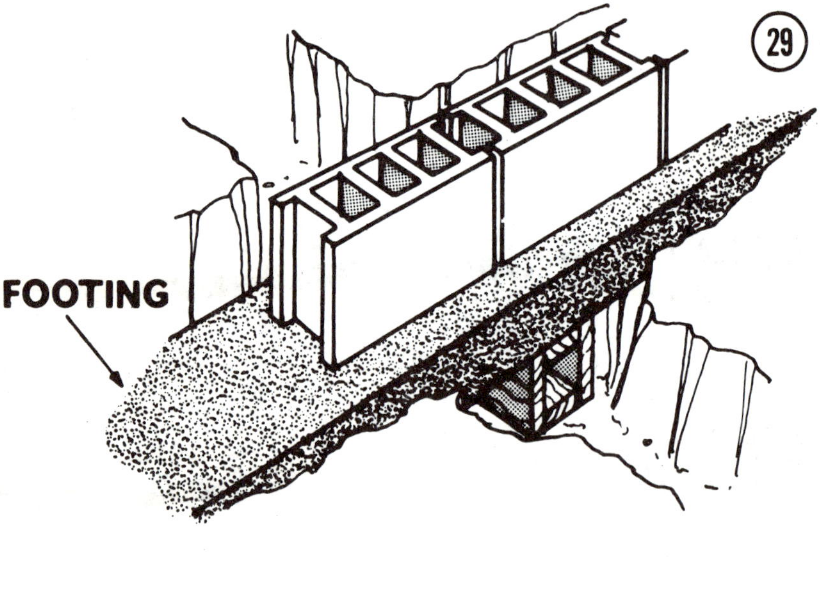

Modifications required to conform to the Canadian Electrical Code -
* See pages 97, 98.

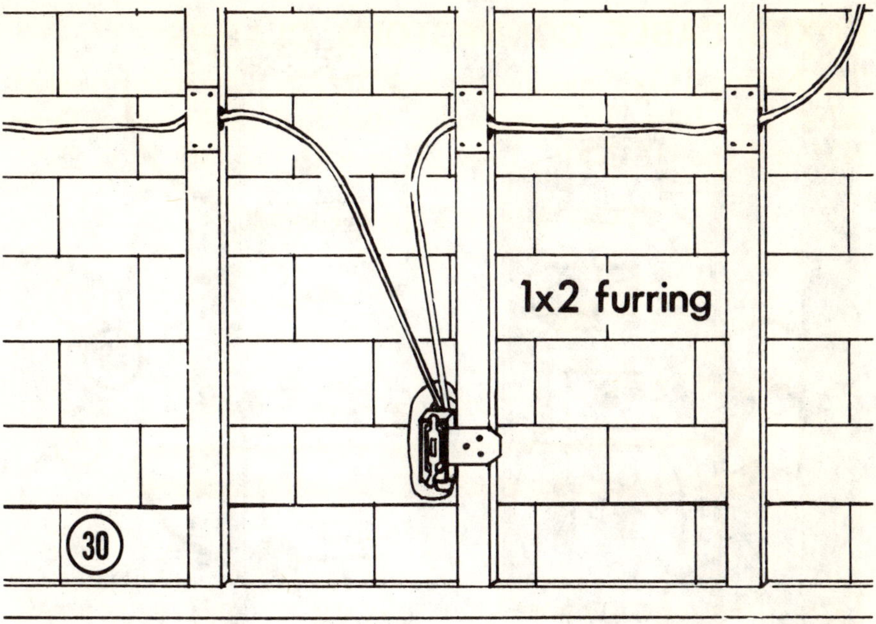

Concrete block houses, Illus. 30, are frequently wired with type NMC. Non-metallic sheathed cable should be fastened every 4½ feet and within 12″ of a receptable box, or distance code specifies. Use straps, Illus. 31.

When installing cable in exposed areas where it can't be supported by existing framing, nail a 1x2, or 5/4x3, by length needed to framing, then fasten cable to 1x2. Always install cable in position where no nails will be driven into it.

Receptacle boxes come in different shapes and sizes. Each can be nailed to studs, Illus. 32, or fastened in position to plaster or wallboard with expansion ears, Illus. 33; mounting hangers, Illus. 34; or fastened to a 1x2 or 2x4 nailed between studs.

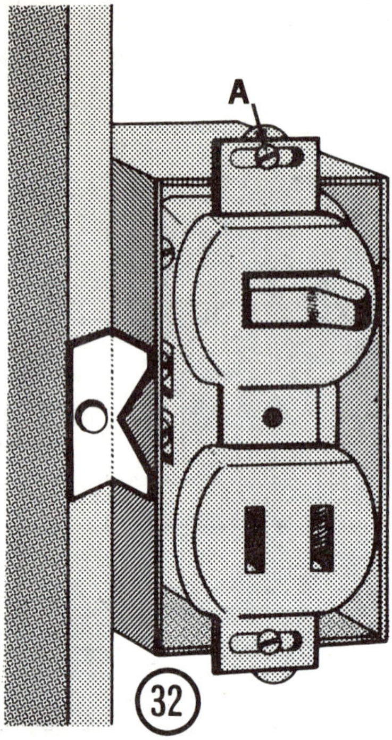

32

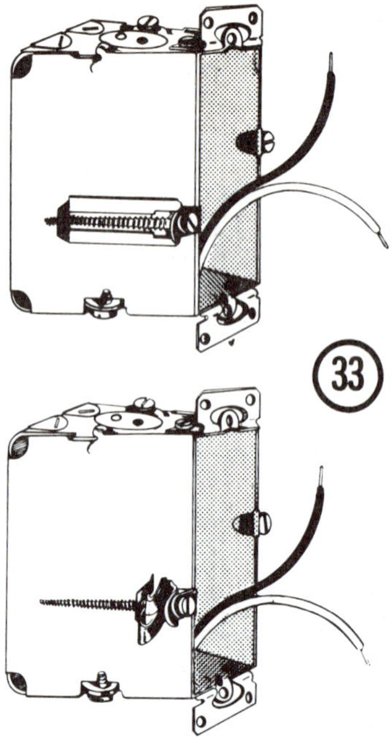

33

To install a box in a gypsum, paneled, or plastered wall, saw hole to size box requires, Illus. 68. Snake BX or armored cable through hole. Fasten connector to cable. Fasten connector to special replacement box, Illus. 33. Place box in opening. Screw in ear expands ear and holds box in position.

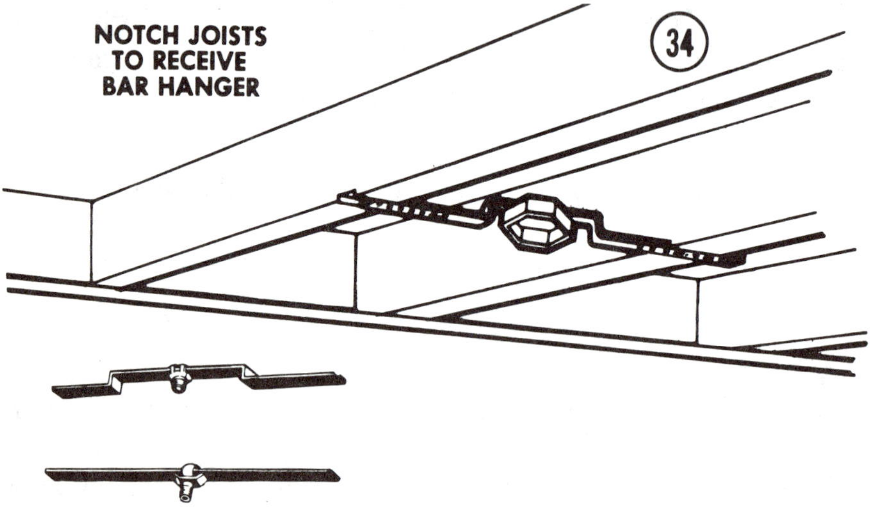

34

Outlet, switch, fixture or junction boxes, all serve one purpose, to provide a safe housing for connections, and for mounting the switch, outlet or fixture. The shape of the box is selected for its end use. Receptacle boxes can be taken apart. Two or more may be assembled together when you want to double up on switches or outlets, Illus. 35. Loosen screw B, remove side and join another box.

35

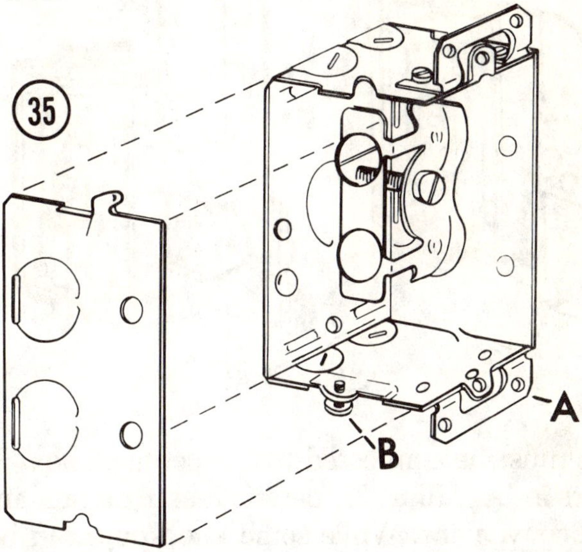

All receptacle boxes should be installed so they finish flush with plaster or wallboard. Where wall covering permits, ears A, Illus. 35, can be screwed to framing with #5 wood screws.

If you want to project box ½″, for ½″ gypsum board, reverse the ears, Illus. 36.

36

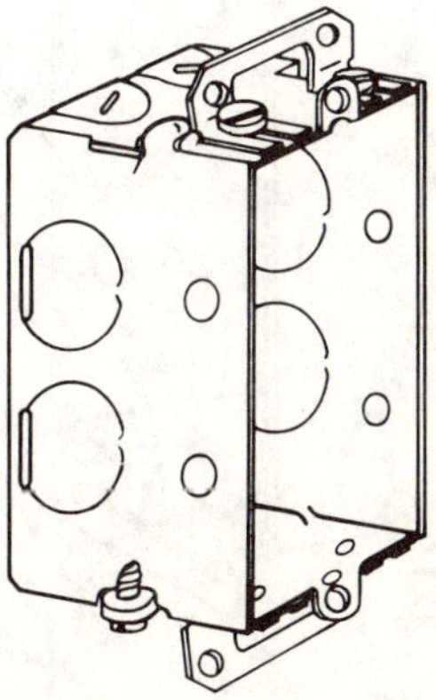

When paneling a wall, you can usually loosen screws A, Illus. 32, to permit switch to finish flush. If you need to apply furring strips, nail boxes to furring strips in position shown, Illus. 37. Boxes now permit fastening to side of stud.

37

All cable must be connected to a receptacle box. Prepunched knockouts in box, Illus. 38, permit making a hole in box where it's most convenient. While some knockouts can be pried out by inserting the tip of a screw driver in slot, knock others out.

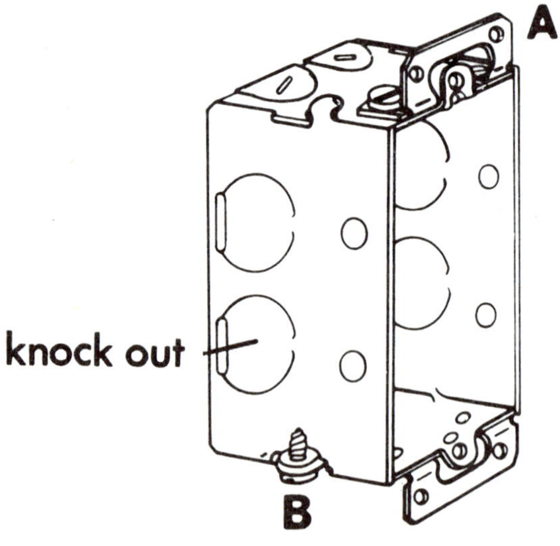

38

Codes permit installing conduit indoors or out. Thin wall conduit, approved by most codes, comes in ten foot lengths. Cut conduit to length required. Use a hacksaw. Always use diameter conduit wire requires. After cutting to length, ream ends using a round file. If you need to bend conduit, use a pipe bender, Illus. 39. These can be rented.

39

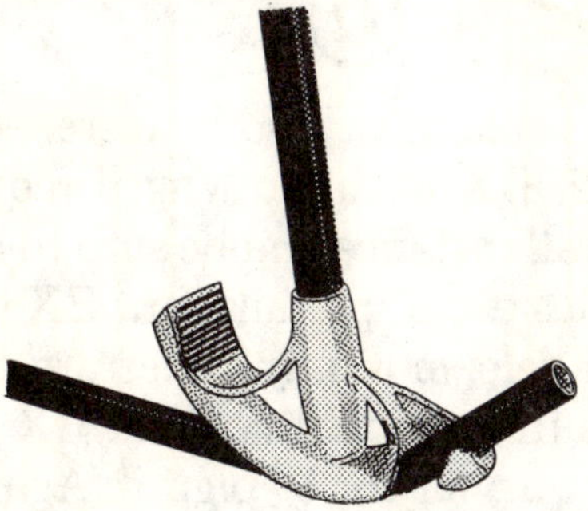

Conduit is fastened to box with a locknut and bushing, Illus. 40. Slip locknut over end of conduit keeping beveled teeth on nut facing box. Place connector in hole. Slide conduit into connector. Slip ring and bushing over pipe and tighten bushing over end of conduit.

40

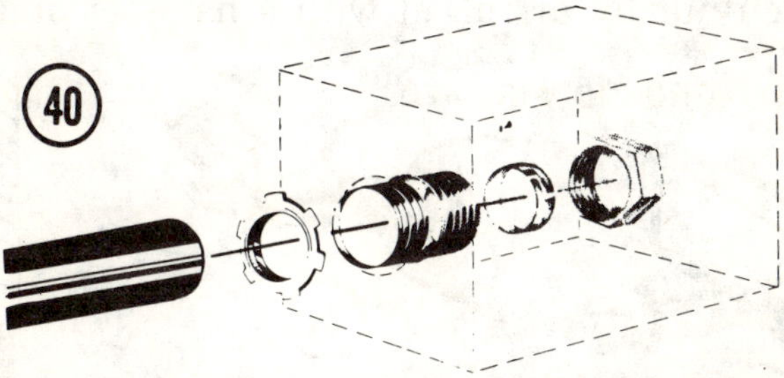

Using a hammer and screw driver, tighten locknut. The rounded inside edge of bushing provides a smooth surface for wires. Cut wires to length required. After conduit is fastened to box, snake wires through conduit. Allow 6″ for pigtails.

Use a coupling, Illus. 41, to join two lengths of conduit. Be sure pressure rings are in position shown.

41

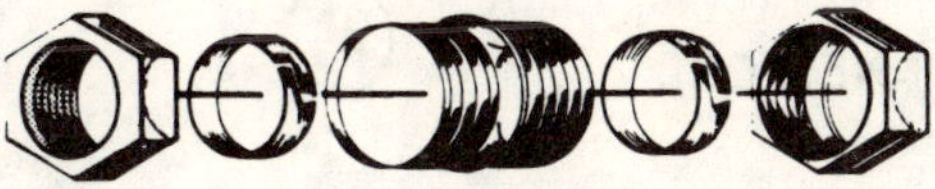

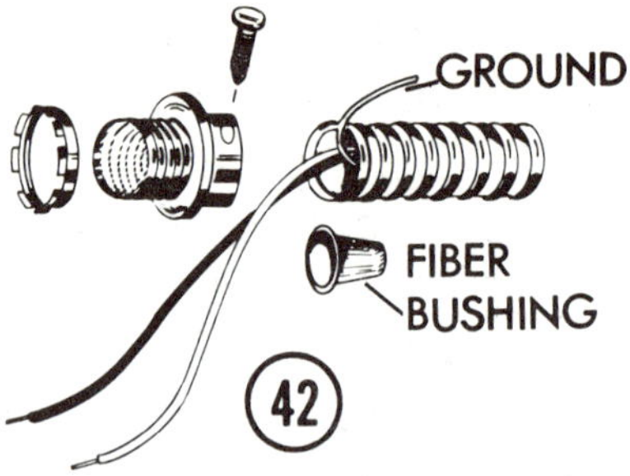

BX and nonmetallic cable must also be fastened to a box with a connector, Illus. 42. BX is cut to length required, plus 12″. Always measure overall distance and be sure to allow for bends. The added 12″ provides two 6″ pigtails. Cut BX with a hacksaw, Illus. 43, at a slight angle, and only penetrate the metal. Stop as soon as blade cuts through metal. Bend BX back and forth and it will break. Cut armor for 6″ pigtail. Again, use care not to cut beyond metal. Remove paper filler. Insert plastic bushing, Illus. 44. Bend ground wire back and wrap it around outside end of metal. This permits ground to make contact with connector, Illus. 45. Fasten connector to BX by tightening screw. Insert pigtails through hole and fasten connector to box with locknut. Tighten nut by driving it with a hammer and screw driver, Illus. 46.

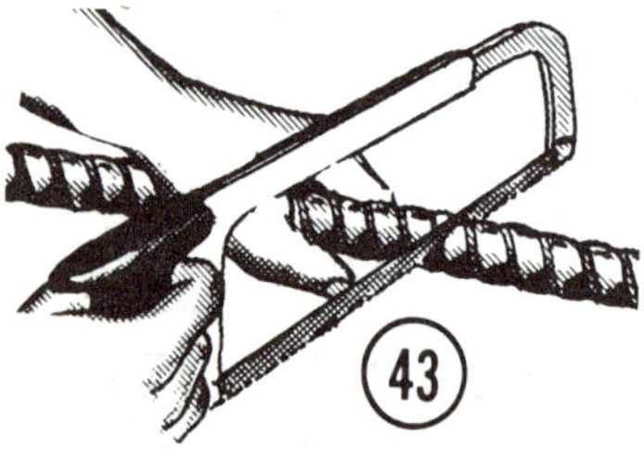

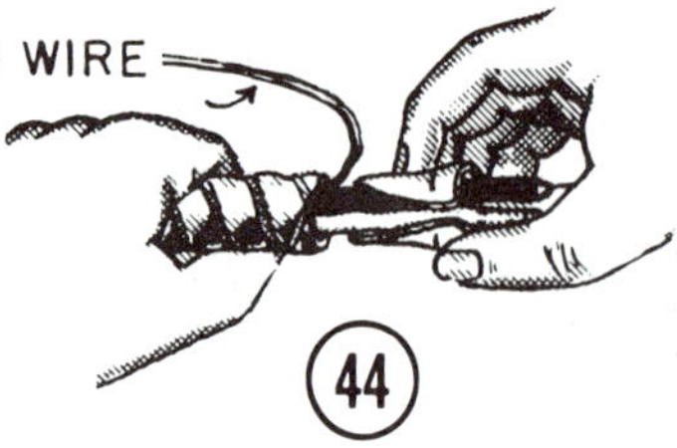

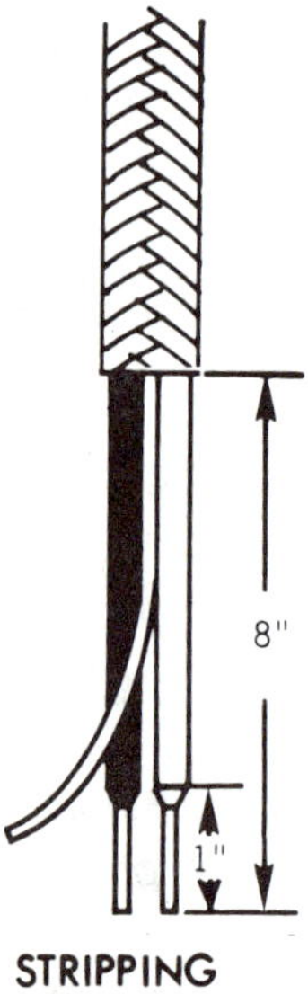

STRIPPING ROMEX CABLE

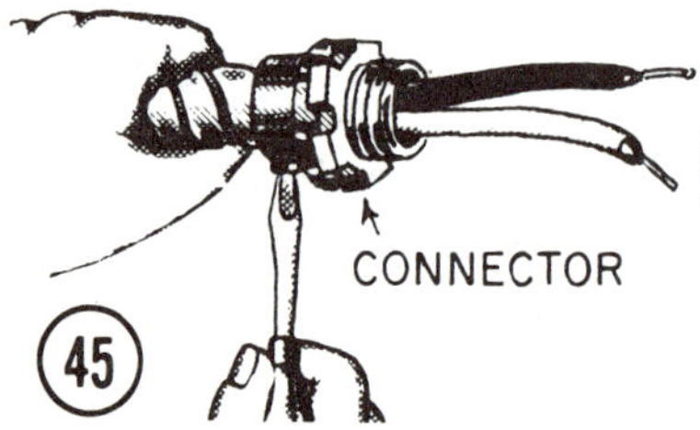

Illus. 47 shows one of the connectors available for nonmetallic cable.

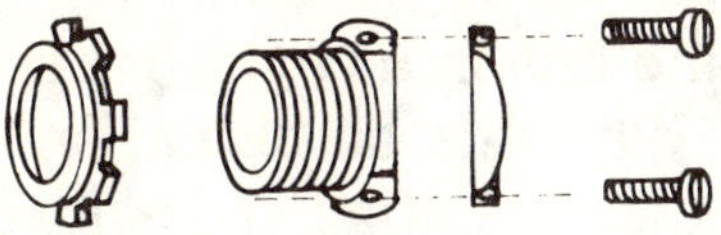

47

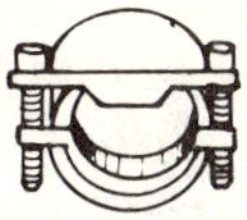

If it's difficult to fasten a connector to outside of a box, use a box with clamps on inside. These can be used with BX or nonmetallic cable, Illus. 35, 48. In this case, remove clamp, pry out knockout using a screw driver, insert cable in box, then fasten clamp over cable.

48

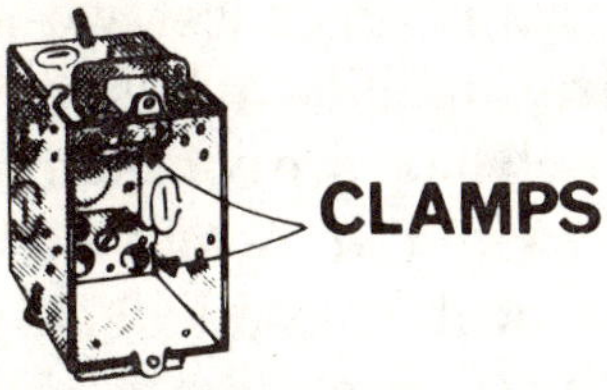

WHEN WHITE IS BLACK

The black wire carries the power. This is rated anywhere from 110, 115, 120 or 125 volts. The white line, always continuous, and always connected to each outlet and fixture, but never to a switch, carries the power back to the source.

If a light fixture was connected to a black and white line without a switch, the bulb would burn continuously, Illus. 49. When a switch is installed and OFF, it interrupts flow of current. Line A, Illus. 50, is continually live, while line AA only comes alive when switch is closed (ON).

49

50

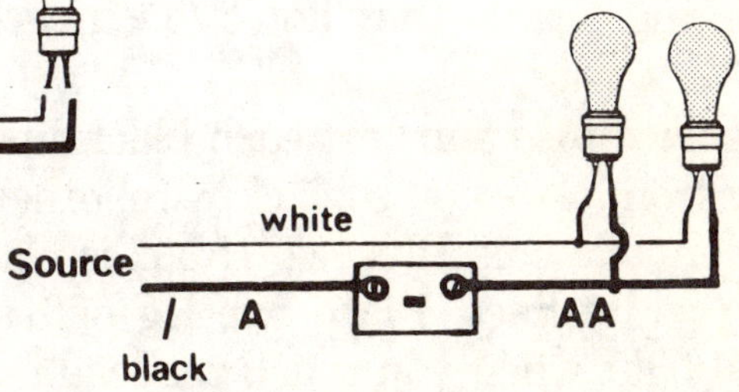

Wiring diagrams usually indicate a heavy black line for the hot, live lead, a thin line for the white, or neutral.

As previously mentioned, the white line in a circuit is always a neutral wire, except where the ends of insulation are painted black, Illus. 24. Codes require painting ends of white wires black when white is used as a live lead. This indicates the white is to be considered a black line. Electricians sometimes neglect to do this, and the homeowner, seeing a white and black wire in a bathroom switch box, begins to wonder.

Illus. 24 shows a typical installation of what is called a switch loop found in many bathrooms. Because two-wire cable contains a black and white wire, and a switch is only installed in a live line, both wires in a switch loop are considered live, (black). Black from source enters switch at A and leaves through B. When you use a tester on a switch, you will only find one lead live when switch is off; both terminals live when switch is on.

When you open a junction or receptacle box and see two or more black wires twisted together, Illus. 22, 23, 6, and protected with a wire nut, only one black will be from source of power, your fuse box; the other blacks carry power to other receptacles.

If you want to install an extra outlet, your first job is to find out which lead comes from source, and is not interrupted enroute because of another switch.

Remove fuse serving circuit. Remove screw and cover plate. Remove screws holding fixture in box. Pull fixture and wires out of box. Remove wire nut on black wires. Separate wires making certain they don't touch each other or anything.

Replace fuse and test each black line with a tester. Place one leg from tester on a black wire, other leg from tester on metal receptacle box, Illus. 22. If box is plastic, remove nut connecting white lines. Place one leg of tester on white. The tester will light when you touch the lead from source.

A wall outlet can be connected to any junction or receptacle box that contains uninterrupted power from source, and a white neutral line. The National Code does not permit removing a single pole switch in a bathroom and replacing it with a combination switch and outlet, Illus. 51, if this switch is connected with a 2 wire switch loop, Illus. 24.

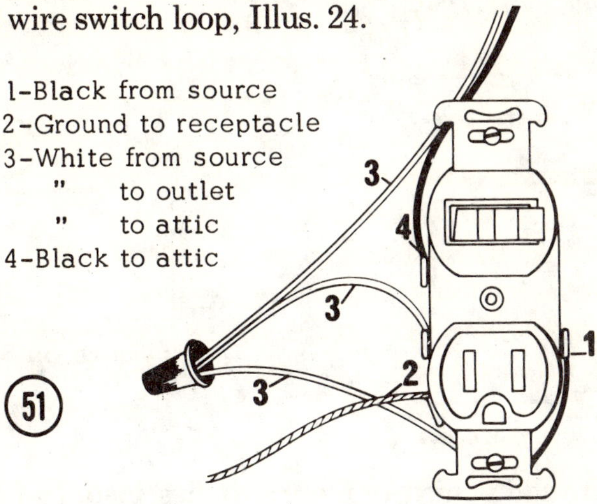

51

Illus. 22 shows a light fixture controlled by single pole switch.

Illus. 98 shows a light controlled by two three-way switches. Note position of fixture in relation to source of power. Black from source is connected to black in two wire cable, and to terminal 1 on switch B.

Using three-wire cable between switches, the red and white wires are used as travelers, and connected in position shown. Contacts 1 and 2 in a three-way switch are permanently connected with a bar, Illus. 52,98.

Note how three-way switches can be switched to different ON and OFF positions, Illus. 52, 53, 54, 55.

The red wire can be connected to 3 on A and B and white between 4A and B, Illus. 98, or as shown in Illus. 52.

Switch is closed (ON) when in position shown, Illus. 53, 54; switch is open (OFF) when in position, Illus. 52.

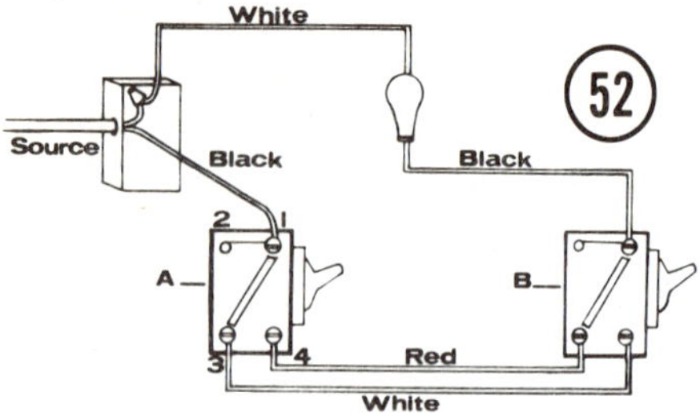

LIGHTS OFF

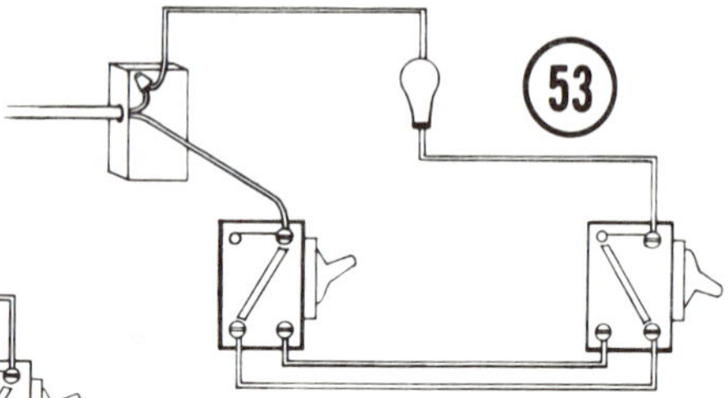

LIGHTS ON

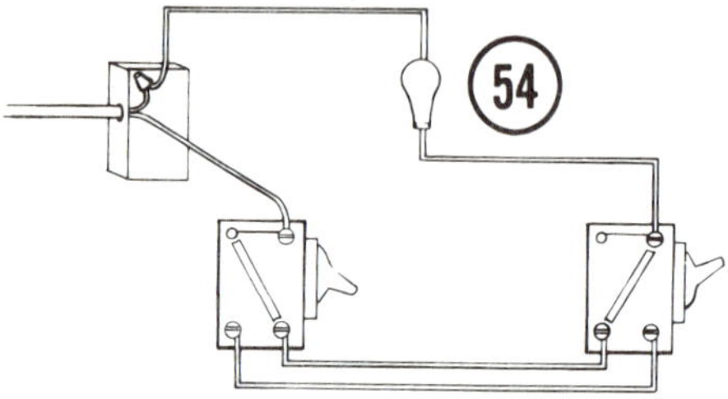

LIGHTS ON

If you want to add a third, or more switches, install a four-way, Illus. 55. This type of four-way switch is connected in position shown.

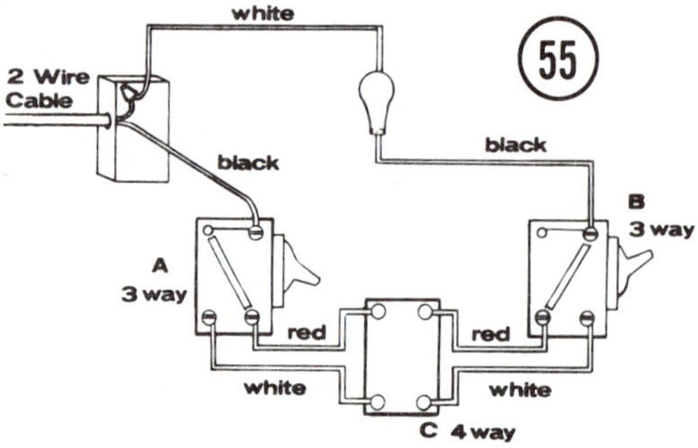

LIGHTS ON

Always install individual 120/240 circuits for a clothes dryer, range, hot water heater, ½- or ¾-ton or larger air conditioner, large motors, water pump, except where other specifications are required.

Always install individual 120 circuits for built-in bathroom heater, dishwasher and waste disposal, workshop, oil burner, ⅓-ton air conditioner, except where other wiring is specified.

Before purchasing new equipment, check "start up" wattage motor driven equipment requires. Decide where it is to be installed. To avoid blackouts, poor TV and radio reception, dimming, or flickering lights when equipment starts, add up present wattage on circuit. Run a new circuit if same is required. Never overload.

GARAGE LIGHTING*

If you want to control garage lighting from the house as well as the garage, use two three-way switches. Note pages 38, 57. Use type NMC 3 wire cable. This can be buried underground.

INSTALLING AN EXTRA OUTLET

If you decide to add an outlet, beyond one that's not controlled by a switch, Illus. 14, the new one will also be continually alive. If you install an outlet beyond one that's controlled by a switch, the new one will be switch controlled.

Any junction box that contains a white neutral, Illus. 14, 22, 23, can provide source of power.

First floor outlets can be connected to any junction box in basement, Illus. 56, or to an existing outlet. Outlets installed in second floor rooms can frequently be connected to a junction box in attic, Illus. 57.

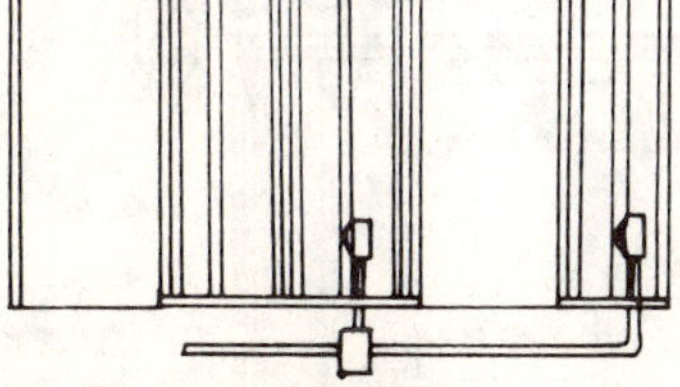

56

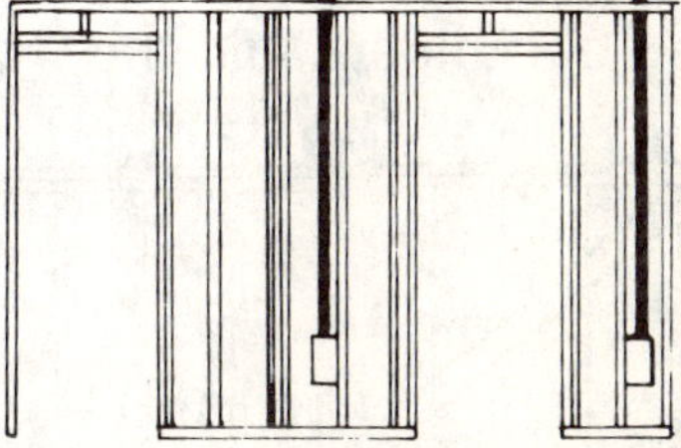

57

* Modifications required to conform to the Canadian Electrical Code - See pages 97, 98.

Always install a new outlet or switch at same height from floor as existing ones. Install between studs in gypsum board, or plastered walls; to studs in new construction.

Follow this general procedure to install an outlet, switch or fixture:

1. Disconnect fuse to power source.
2. Select location, in new construction, fasten box in position.
3. Locate a power source, another outlet, junction box, etc.
4. Cut opening for box (when remodeling).
5. Drill holes required to run cable.
6. Remove knockout in position cable requires.
7. Strip cable to allow a 6″ pigtail if BX or nonmetallic cable is used; 8″ if wires are run thru conduit.
8. Snake cable between opening and source.
9. Insert fiber bushing if BX is used.
10. Fasten connectors to cable.
11. Insert pigtails thru hole in box.
12. When remodeling fasten connectors to box, box in position.
13. Connect pigtails to switch, outlet, or fixture.
14. Connect pigtails to source.
15. Replace fuse.

Wall studs in older houses were frequently spaced 18″ or 24″ on centers. During the last 30 years, most houses were built with studs 16″ on centers.

An outside wall frame, Illus. 58, that contains a picture or a bow window will usually be framed with a double 2x4 plate, and double headers over windows and doors. Window sills will be a single or double 2x4. House framing varies with local codes, different builders, and the trend at time house was constructed.

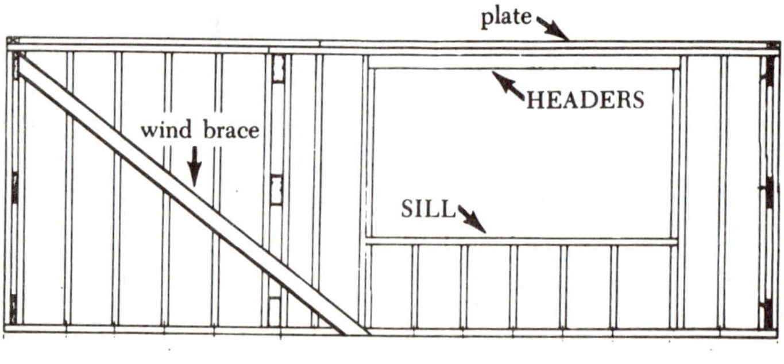

58

Interior doors are usually framed with a double 2x4 or 2x6 header on edge, Illus. 59. Whenever interior walls butt together you invariably find 2x4 spacer blocks A. These will be 1⅝″, when edgewise; 3⅝″ — B when flatwise. If 2x6 is used, it measures 5⅝″ when used flatwise.

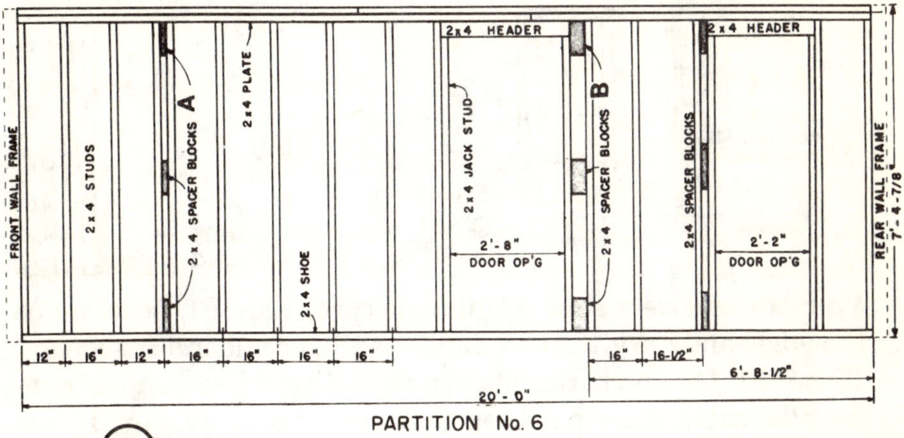

59

Partitions are frequently stiffened with cats, Illus. 60. Use 1x2 or 2x4 if you want to install a box between studs. Position framing so front edge of box finishes flush with plaster or wallboard.

60

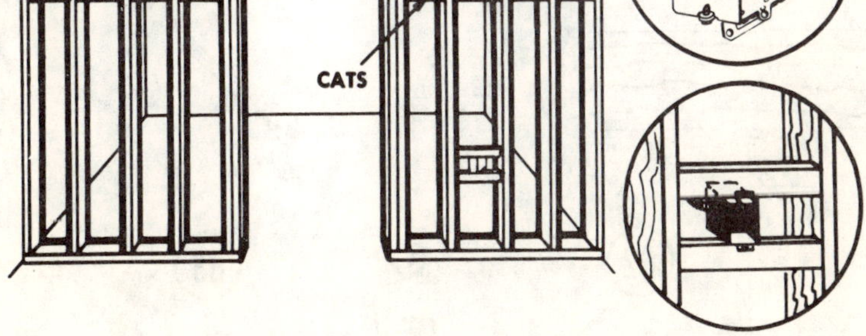

To simplify installing a new outlet, always try to connect it to an existing one. This can be done in a number of different ways.

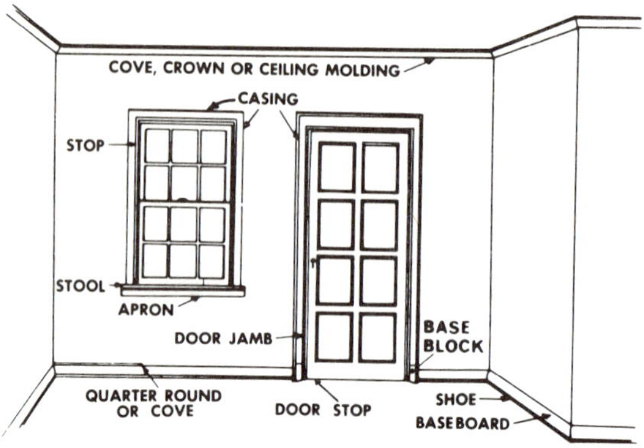

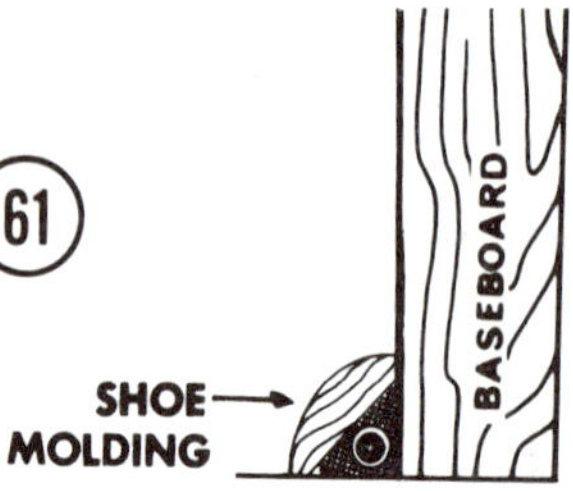

61

You can remove a shoe molding or cove, Illus. 61, bevel plane the edge, and bury the wire underneath, or under a baseboard, Illus. 123. Or notch the plaster or wallboard to receive non-metallic cable. New patching cements simplify repairs. If you have to run a line into the basement, remove the shoe molding and use a ⅝″ or ¾″ spade bit, Illus. 62, fastened to an 18″ or 24″ extension.

New tools now simplify running wires in the easiest possible manner. These consist of a ⅜″ x 54″ long spring steel flexible shaft drill, Illus. 63, an alignment guide, plus a wire gripper. The gripper permits using the drill as a snake.

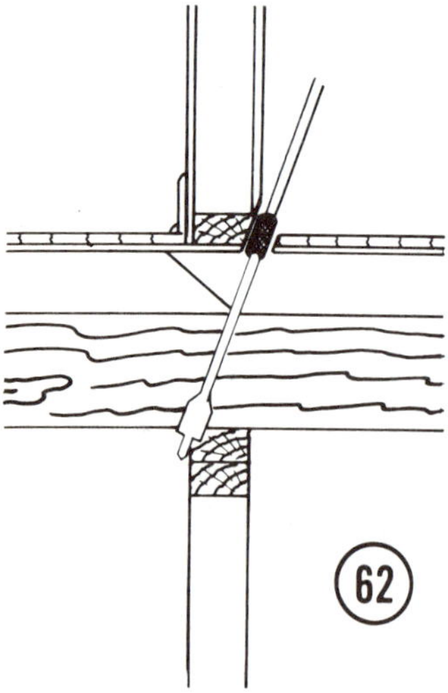

62

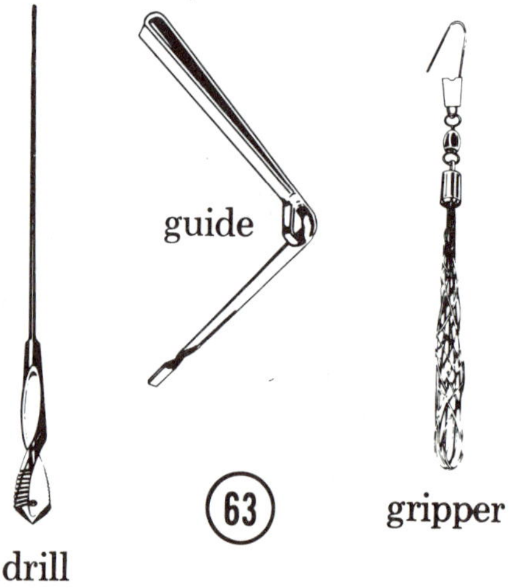

63

To fully appreciate how easy this can be, note Illus. 64. After removing shoe or ceiling molding, or after selecting a location for an outlet, drill a 1″ hole, in position required, for the alignment guide. The guide is inserted in the hole. The ⅜″ x 54″ bit is inserted in a variable speed drill. The bit goes through hole in guide and is guided to arc through a sill or plate into the basement or attic. The 54″ length permits drilling clear through. The spring steel permits keeping the arc while drilling. Since you drill at a slow speed, 850 RPM, you can maintain the position of the bit despite the arc.

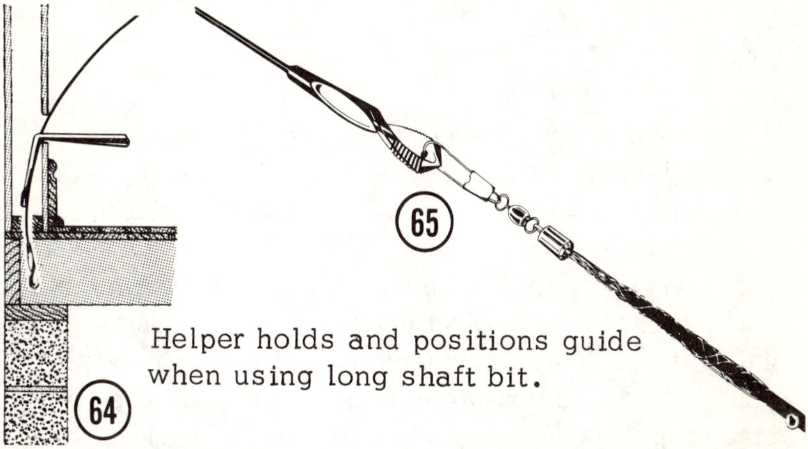

Helper holds and positions guide when using long shaft bit.

When you have drilled clear through, merely fasten the safety pin on end of gripper onto the end of the bit, Illus. 65; insert wire into gripper, draw bit back up through the guide.

Use a variable speed drill (850 RPM to 1500) with a reverse gear. While the average homeowner may not want to invest money into a set of these tools for one job, even though the savings of doing one job may pay all costs, tool rental stores have these tools available.

When you want to run a cable through a wall from room above, the long bit simplifies the job. If you measure distance from an outside wall you can usually cut opening for box in direct line with hole. If you run into a cat halfway between, chip the plaster and recess the cable, Illus. 66. In new construction always cover notch with a steel plate.

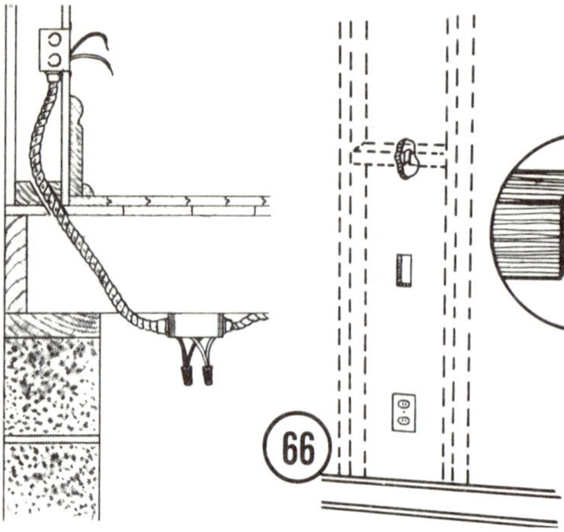

66

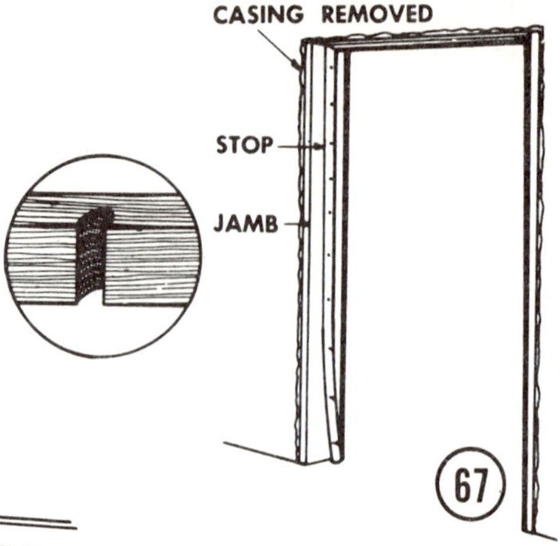

67

If you want to install an outlet beyond a door, and can't conveniently connect to a junction box in basement, you can plant cable under baseboard and under door casing. Illus. 67 shows casing removed from around an interior door. This is usually nailed with 8 penny finishing nails. To remove, start at base block, Illus. 61, and carefully pry it loose. Using a pry bar, placed under center of bottom end of casing, carefully pry up casing. When side casing is removed, pry up top casing then remaining side. Notch plaster or wallboard close to jamb to depth cable requires. Fasten cable in recess, then replace casing. Countersink nails, fill holes with putty or wood filler, then repaint casing.

When planning new outlets, always make a survey. For example, if you install an outlet below a picture window, or in a wall opposite an existing outlet, consider whether it's easier to make a connection to a junction box in basement, or from an existing outlet, Illus. 56, 57, rather than go around a door. Make certain junction box selected can take extra wires.

If you are installing a box in a lath and plaster wall, punch a small hole in position selected. Using a screw driver, chisel a narrow vertical slot to expose width of one lath. Use the full size template, Illus. 68, to draw outline of box. Position template so you only saw and remove one lath, and only notch the lath above and below.

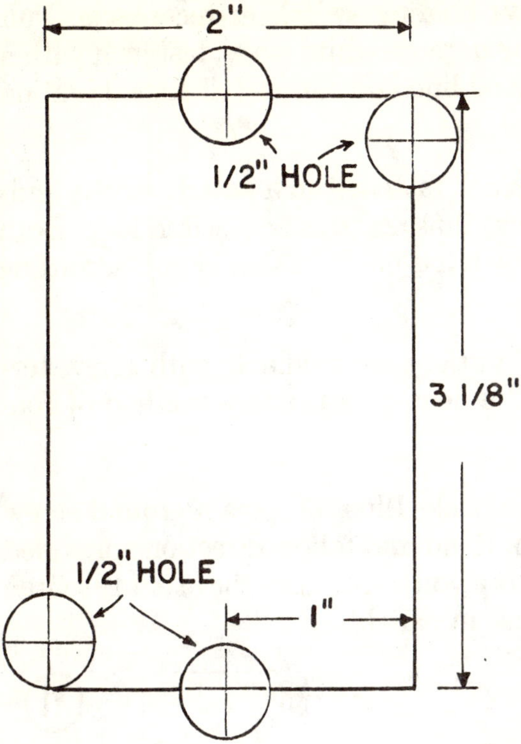

68

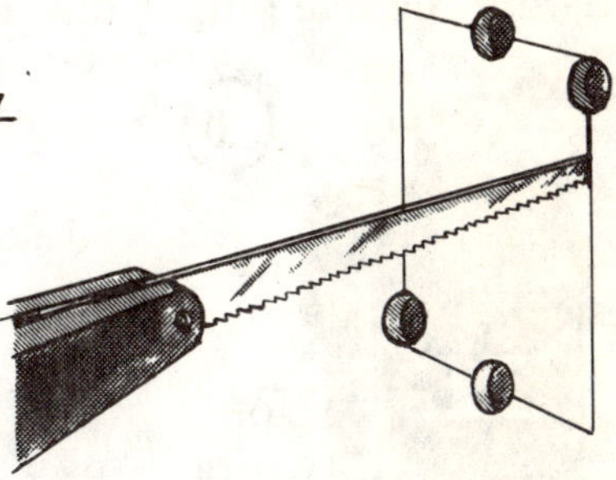

69

While this may alter exact floor height of box, no one will know if you don't tell them. Drill ½-inch holes in position noted, Illus. 69. Saw plaster along line of template with a metal cutting keyhole saw, or use a hacksaw blade. Place a small block of wood alongside line of opening and hold saw blade so it only cuts when pulled toward you. If you press block when you pull blade, you can make a clean cut. If you happen to tear any plaster off wall, same can be repaired with patching plaster after box has been fastened in place.

If you need an extra outlet, and want to make a quick installation by removing a single pole switch to install a combination switch and outlet, Illus. 51, it can be done if you find a neutral wire in box, as shown in Illus. 22, 23.

Remove plate and screws holding switch in receptacle. Pull out switch. If you find one or more white wires, fastened with a wire nut, you have a neutral line and can install a combination switch and outlet.

If you find a white and black fastened to a switch, or the ends of the white painted black, Illus. 24, this is a switch loop. Both wires will test hot when switch is on. You should not, according to code, install an outlet.

Combination switch and outlets are available with screw terminals, Illus. 70, or use the new, pressure lock method of connecting wires.

These have three holes in back, Illus. 71, plus a ground screw connection painted green. Read and follow directions provided by manufacturer. But before you begin, turn the existing switch on, and remove proper fuse in panel.

(70)

Note wiring Illus. 51

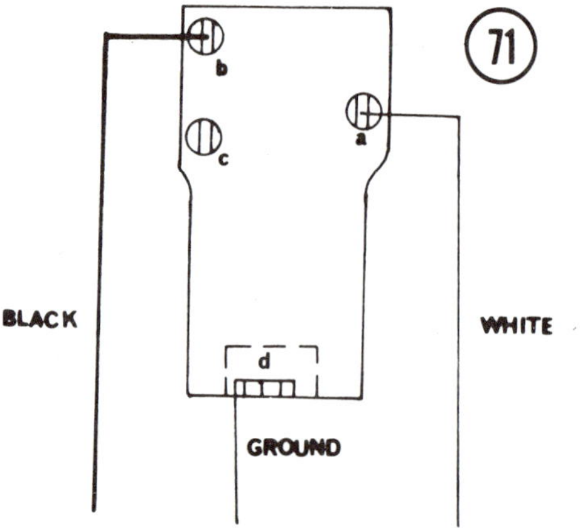

(71)

Remove screw and cover plate. Remove screws and pull switch out of box. Remove wires from existing switch and if you use a pressure lock outlet, straighten bare wire ½″. Be sure to measure and cut it exactly ½″.

The replacement outlet can be connected in 3 different ways.

Illus. 71 shows a switch controlled outlet independent of light fixture.

Illus. 72 shows a switch controlled outlet that works only when light is on.

Illus. 73 shows how to connect a continuously live outlet while the light can be switched on or off.

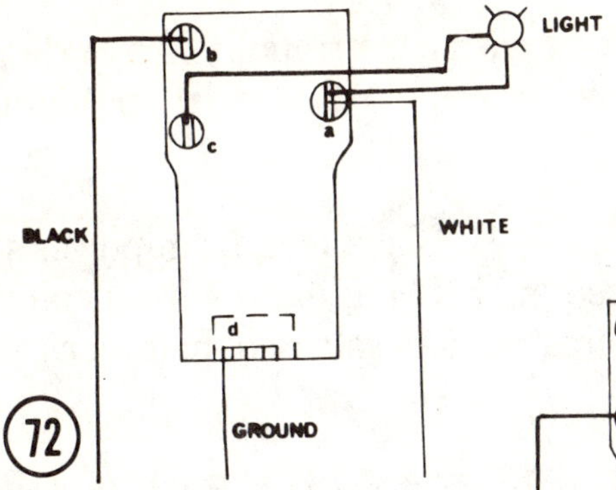

72

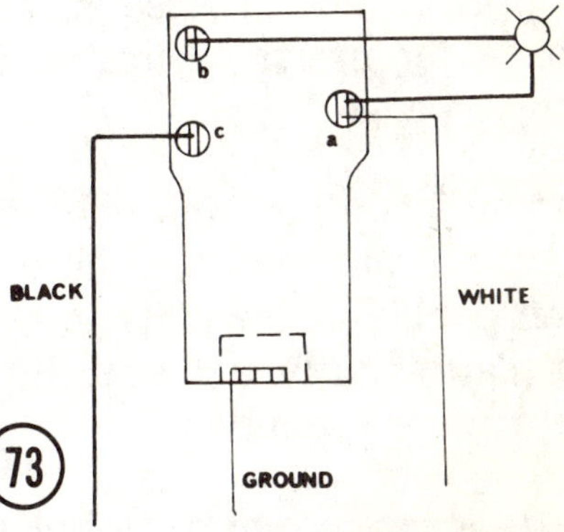

73

Always strip ends of wire ½″ when inserting one wire in a hole.

If you want to disconnect a wire after insertion, press a small screw driver, one not wider than ⅛″, into slot next to hole. Press screw driver down and pull wire to release.

To connect a switch controlled outlet, Illus. 71, cut a 6″ length of white wire, strip ends ½″ and insert one end in A. Connect other end to white neutral. Cover with a wire nut. Cut a piece of black wire, strip ends ½″ and insert one end in B. Connect other end to black. This outlet will take a plug with a grounding prong, Illus. 17.

To make a ground connection, cut a piece of #14 single strand wire and strip insulation ⅝″ from end. Slide this end through clip, Illus. 16. Make an eye on other end, and fasten to green ground screw.*

*Modifications required to conform to the Canadian Electrical Code - See pages 97, 98.

For switch controlled outlet and light, Illus. 72, cut and strip white wire 11/16". Do the same with white lead from light fixture. Insert both into A. Strip black wire from light ½" and insert into C. Strip end of black line from source ½" and insert into B. This hookup provides a switch controlled outlet and light.

For a switch controlled light with a continually live outlet, strip two white leads 11/16" and insert, do not twist, both into A, Illus. 73.

Black from light is inserted in B. Black lead from power source is stripped ½" and inserted in C. The ground wire is connected to box. After inserting wires in holes selected, hammer grounding clip on edge of box.

Insert outlet in box, and only start screws. Replace fuse and test switch and outlet. If all is OK, fasten screws in place. Fasten double outlet plate in position. If edge of clip offsets cover plate slightly, don't let this annoy you. It actually helps make a better ground.

Illus. 74 shows a Safety Outlet with an automatic snap cap. This is designed for families that have children prone to push hairpins and small articles into outlets. The cover is spring loaded so it must be turned to insert plug. While not guaranteed to mystify a mechanically inclined youngster, it does provide some protection. It is installed in same manner as other outlets.

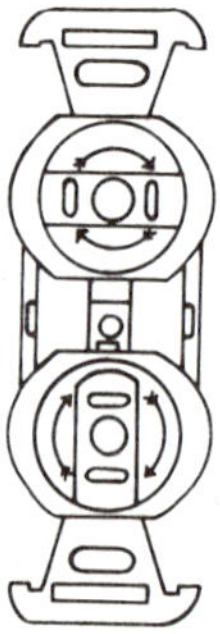

74

WEATHERPROOF OUTLET

An outside weatherproof outlet, Illus. 75, can be installed with very little effort, if you connect same to an inside box. Disconnect fuse and remove outlet from box, Illus. 76. Using a long shank screw driver and hammer, drive out knockout plate in back of box, Illus. 76.

75

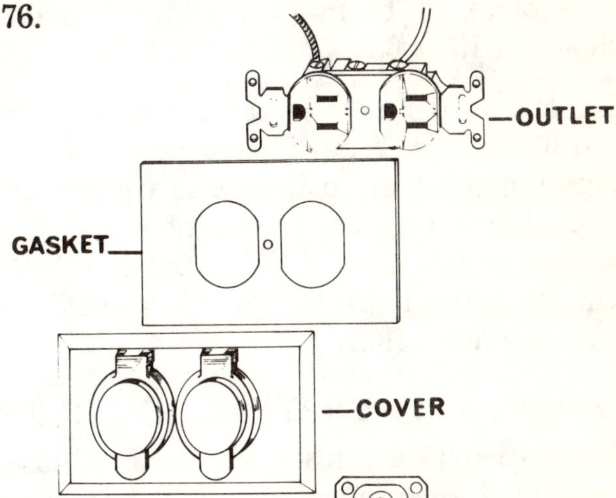

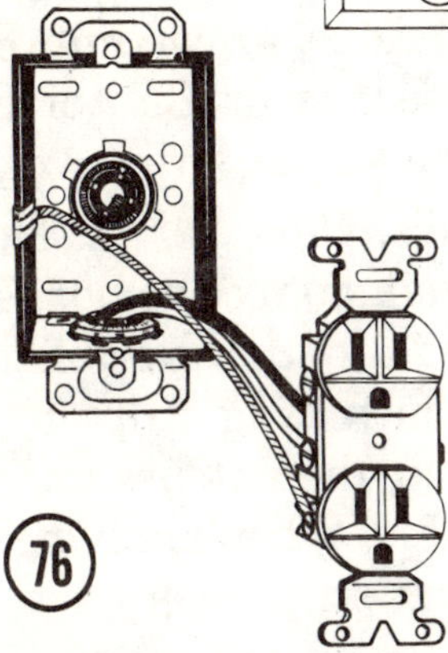

76

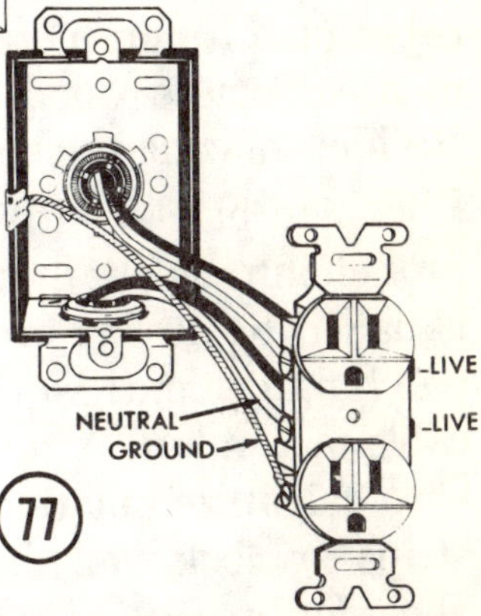

77

Always install an outside weatherproof outlet above point where snow may build up. Install box so weatherproof cover can be installed horizontally, if same is designed for that kind of installation.

To cut hole to exact shape receptacle box requires; use the full size pattern, Illus. 68. Always cut opening in clapboard and sheathing to shape box requires.

Remove knockout plate in box in position cable requires.

If you are installing an outside weatherproof outlet where it's inconvenient to make a back-to-back installation, position new outlet between studs. Measure distance from end of building. Go into basement and measure same distance. Move location if you run into any obstruction. Most houses are framed with a bedplate, sill beam, floor joist, sub and finished flooring as shown in Illus. 63.

Cut BX or non-metallic cable to length required, plus 12″ for two 6″ pigtails. Wrap ground wire around end of insulation, slip connector in position and tighten screw. Slip pigtails through hole and fasten connector to box. Connect pigtails to interior outlet, Illus. 77. Screw outside box to sheathing, fasten wires to outlet, outlet to box. Place gasket in position, screw cover to outlet, Illus. 75.*

WIRING—CONCRETE BLOCK CONSTRUCTION

The National Electrical Code specifies house wiring be armored cable (BX), non-metallic sheathed cable, or conduit. Builders in areas where wood framing is used, favor BX or conduit. Builders of concrete block houses nail furring strips, 1x2, 1x3, or 5/4x3, to blocks with steel cut nails, Illus. 30, 78. Holes are broken into blocks in walls above grade to receive box. The box is nailed to furring and the opening around box is filled with patching cement. Notches are cut in furring strips to receive cable or conduit. After running cable, the notches are covered with galvanized plates to protect cable. When applying furring strips to block walls below grade, cut strips every two feet to allow for breathers.

78

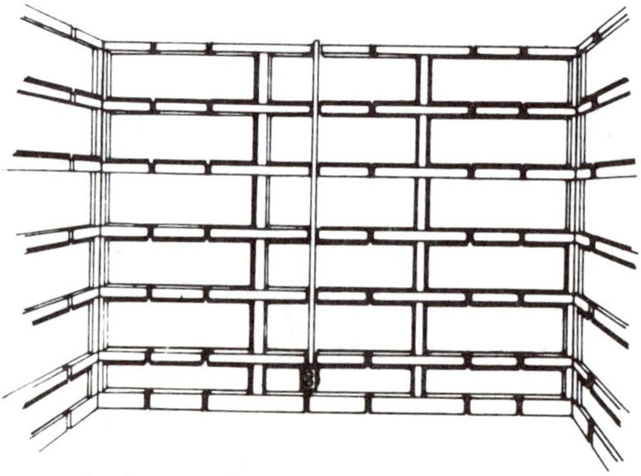

* Modifications required to conform to the Canadian Electrical Code - See pages 97, 98.

TO REPLACE A FRAYED LAMP CORD

Pull plug from outlet. Purchase a 6 ft. or 9 ft. replacement cord in either brown or ivory, Illus. 79. These have one free end, stripped and ready to be fastened to interior C, a molded plug on other end. Remove shade, shade holder and bulb. Loosen screw in Cap A, if lampholder has one. Using a small screw driver, pry cap from shell. Disconnect wires. Disconnect or cut plug off old cord, Illus. 80. Fasten free end of new cord to plug end of old cord and carefully pull new cord through lamp by pulling old cord at top. Connect new cord to interior C. Reassemble lamp. If interior needs replacement, buy identical replacement, Illus. 81.

79

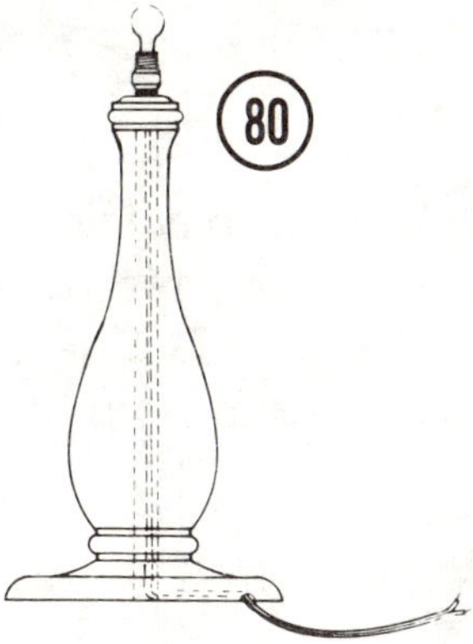

80

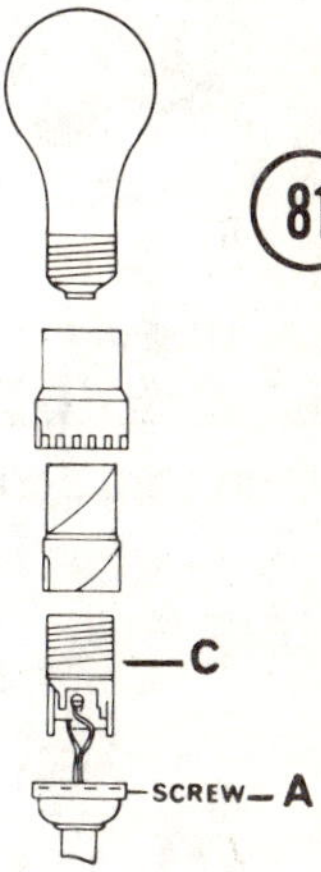

81

MAKE A NIGHT READING LAMP

If you like to read or write in bed and don't want the light to disturb anyone, plug a night light into an extension cord, Illus. 82. Switch on night light eliminates need to leave bed. It can even be used under the covers when you just have to finish a chapter after the windows are open on a cold winter's night. A perfect gift for the cooperative hospital patient who needs to share a room.

82

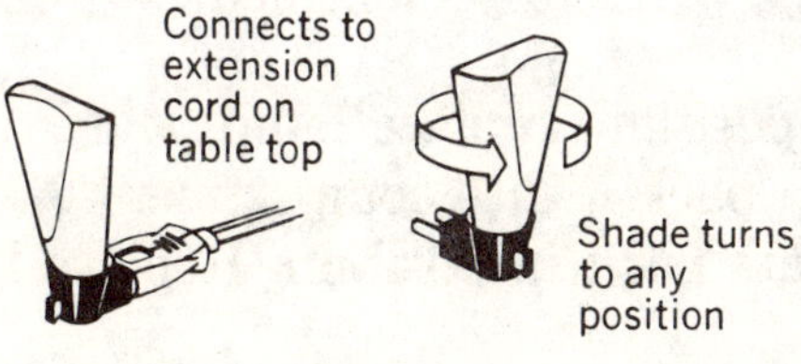

MEDICINE CHEST REPAIR

In time the canopy type switch, Illus. 83, in a medicine chest will require replacement. To obtain correct size canopy switch, loosen nut A, Illus. 84, and purchase new switch that fits nut. Be sure to test nut on new switch before purchase to make certain you have the proper size.

83

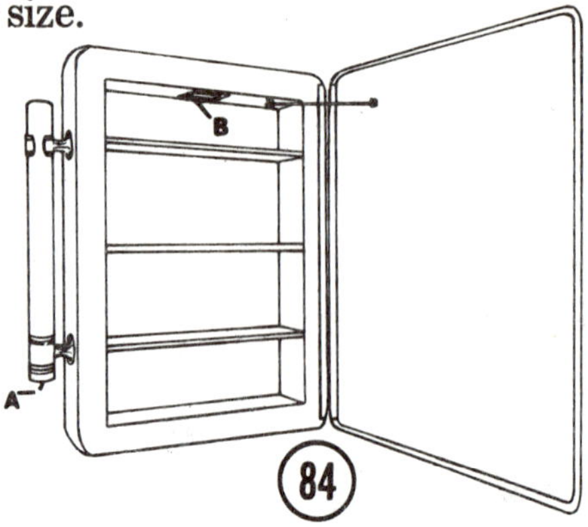

84

Remove fuse. The wiring in most medicine chests is joined in compartment B, Illus. 84. Remove screws and inspection plate. Pull down and straighten wires, but DO NOT DISCONNECT, Illus. 85. Loosen screws and remove lamp bracket. Pull canopy switch out.

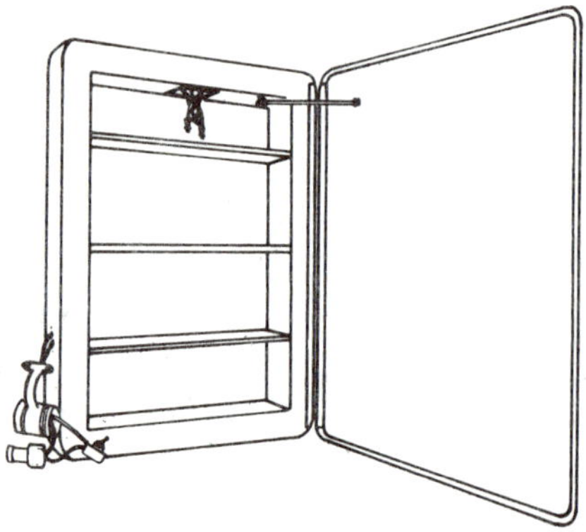

85

Since wires from lamp holder are frequently soldered to switch wires, cut wire off as close to old switch as possible. Strip wires ½″, twist and tape wires to replacement switch.

Insert switch in position, replace lamp holder in fixture, pull slack wire up and pack it into compartment A. Fasten nut to switch. Screw plate back in position. Replace light bulb. Replace fuse.

ELECTRIC RANGE

Connecting a new range to an existing range outlet requires a male plug, Illus. 86,* that matches your female outlet. It also requires connecting three wires to three terminals on range. Each is clearly marked. We don't recommend using an existing cord as it may have been subjected to considerable use and misuse. Twist, and/or, pull plug out of wall outlet, then remove plate covering terminals of range. Remove old cord.

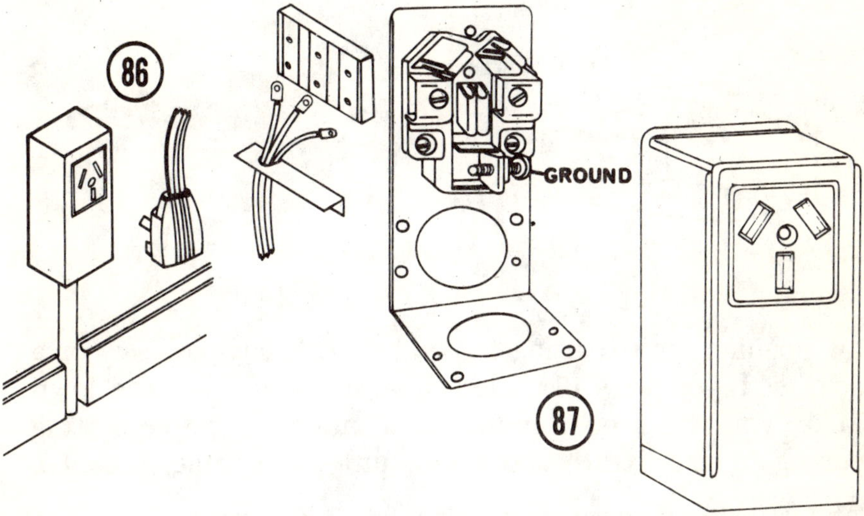

Range terminals consist of two hot and one neutral. The hot leads on power cord are usually brass, the neutral silver. Connect wires to three clearly marked terminals. Replace cover. Push plug into wall outlet. Replace fuse and test range. Always check plate on range and use size of wire range manufacturer specifies.

If you happen to have an odd shaped female outlet and can't locate an extension cord with a plug that matches, pull fuse on main switch. Remove and replace outlet with a matching extension cord, Illus. 87.*

If you need to install a new circuit for a range, use three wire #6, and fuse it with a 50 amp fuse.

*Modifications required to conform to the Canadian Electrical Code - See pages 97, 98.

TO INSTALL A CEILING FIXTURE

To install a recessed ceiling fixture in a gypsum board or plaster ceiling, follow this procedure. Select a location between ceiling joists. These are usually spaced 16″ on centers. Place the fixture in position selected, Illus. 88, and draw outline of box; do not include the wiring receptacle. Draw diagonal lines and make a small hole at center. Bend a piece of coat wire hanger at right angle, push it into hole to see if test-hole is equal distance from joists. If necessary, redraw outline of box to place it in the clear.

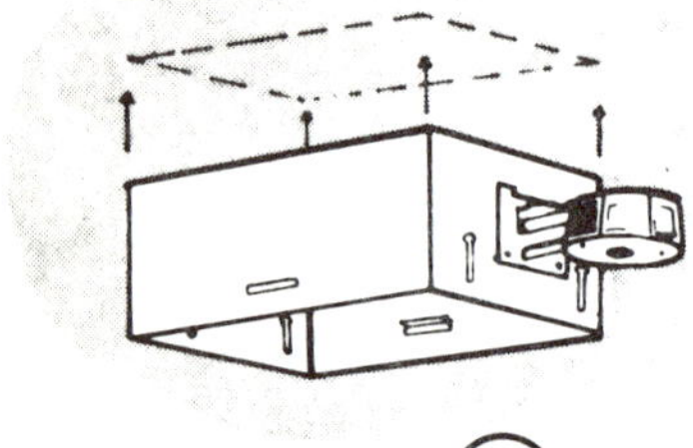

88

89

Saw opening for box using a metal cutting compass saw or a hacksaw. If you press a 1x2 along line, and only cut on the down stroke, you will keep chipping to a minimum. After cutting opening to size fixture housing requires, run cable, Illus. 89.

Connect cable to receptacle, Illus. 90. Slide receptacle end in first. Since this is placed parallel to joists you will have ample room to ease the box in place.

90

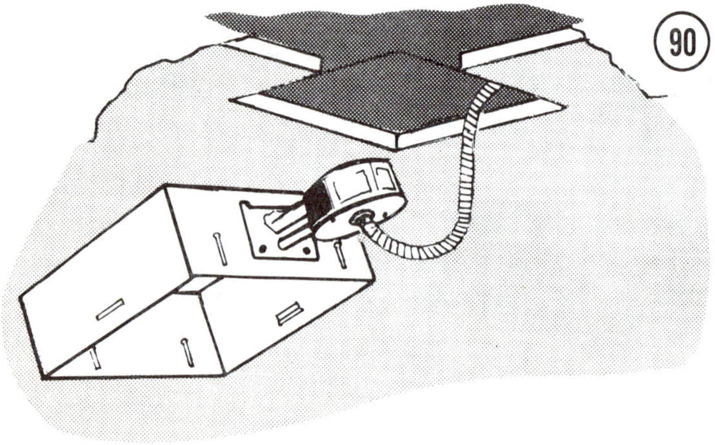

Fasten fixture to ceiling clip, Illus. 91. Install bulb and fixture cover, Illus. 92.

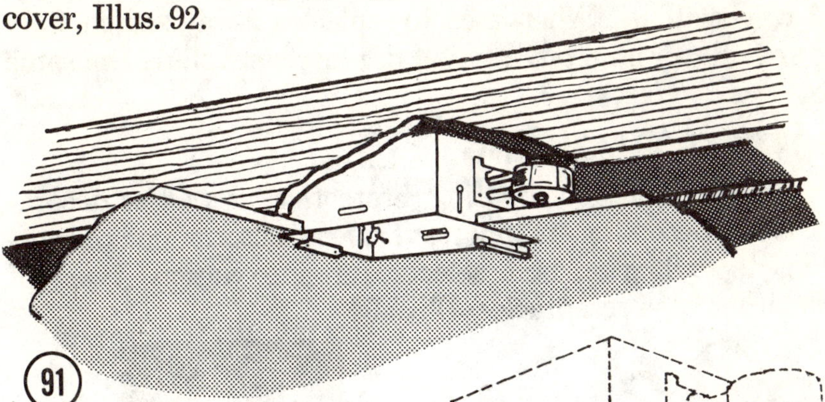

91

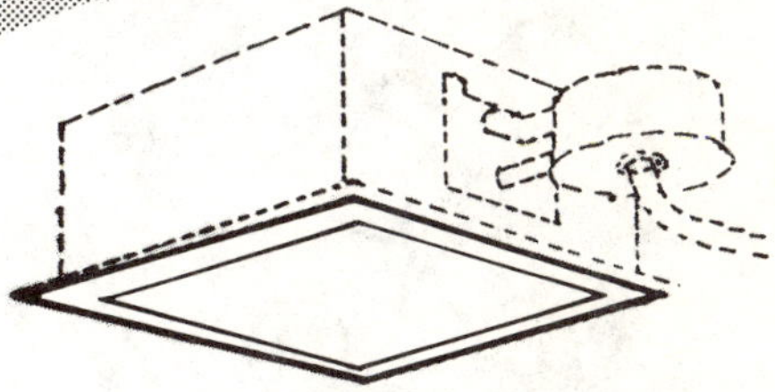

92

If you are installing a recessed fixture in new construction, always nail 2x4 or 2x6 cats between joists so fixture can be fastened to cats. Edge of fixture projects below edge of joist a distance equal to thickness of ceiling material.

If you want to relocate an existing ceiling fixture, connect cable required for new installation in a junction box, and cover box with a box cover.

HOW TO INSTALL DIMMER SWITCH

Dimming a light so it complements TV viewing is extremely beneficial since it lessens eye strain and glare. Built-in fixtures, controlled by a single pole switch can be converted into dimmers. Flip switch on and remove proper fuse. Remove screws holding cover. Pull out switch. Remove wire nuts and connect new dimmer, Illus. 93, white to white, black to black. Replace wire nuts. Replace switch plate. Press on knob. Replace fuse. Push dimmer for ON - OFF. Turn to dim.

93

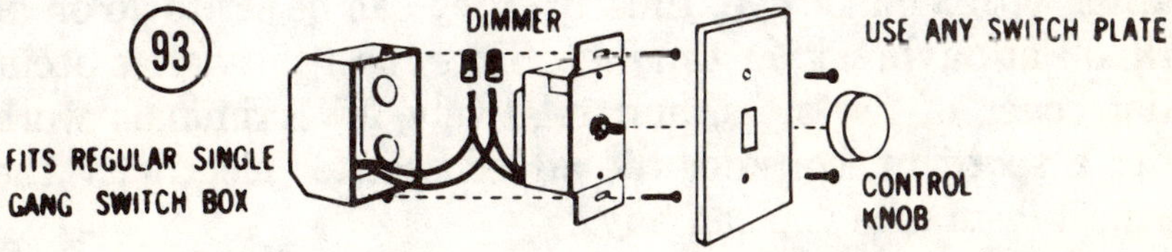

Dimmer switches are available for single pole switches as well as 3-way switches. When used to replace a 3-way switch, only one switch can be replaced by a dimmer controlling the same light.

PROTECTIVE LIGHTING

Installation of outdoor lighting, protective, as well as decorative, is fast becoming an important safety factor. Flood or spotlights, depending on bulb selected, fit into various types of all-weather base receptacles, Illus. 94.

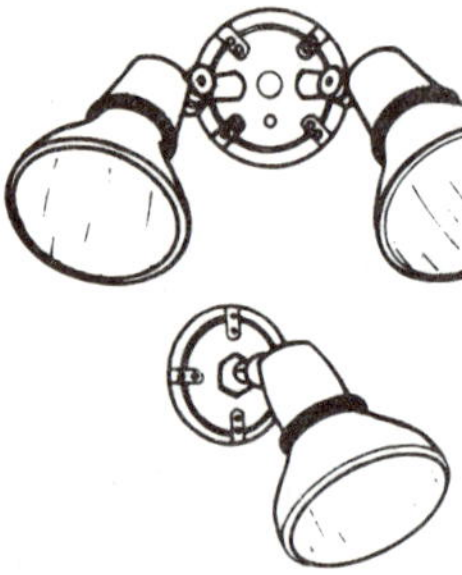

94

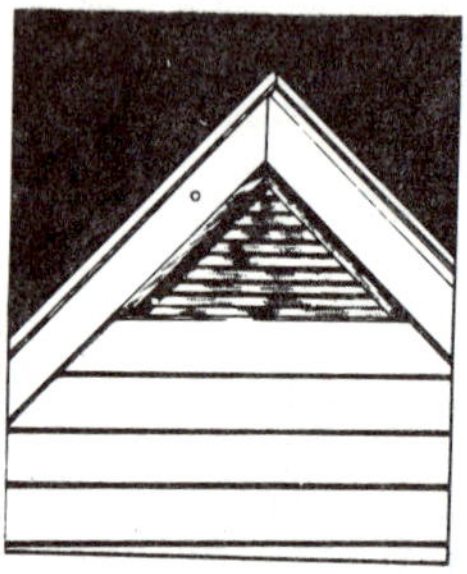

95

These should be fastened to fascia, as high up as possible, on gable ends, Illus. 95.

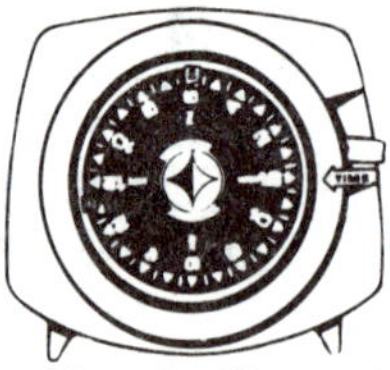

Plug In Timer

AUTOMATIC TIMER

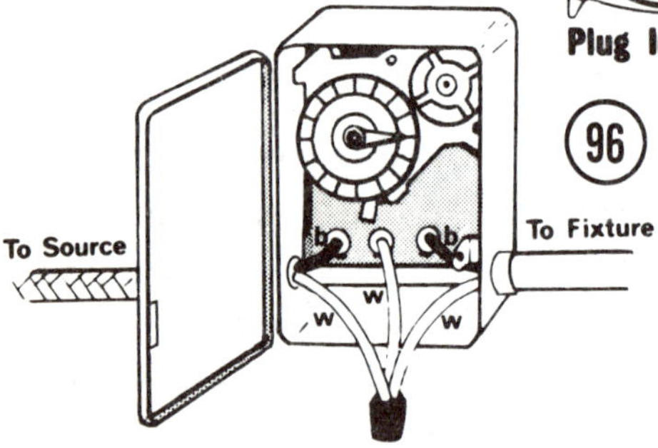

96

97

When connected to an automatic timer, Illus. 96, or a bulb adapter with a magic eye, Illus. 97, they can be set to go on at dark, off at daybreak, or time set. Since most burglaries occur under cover of darkness, and relatively few criminals work under a spotlight, consider the added cost for electricity the cheapest insurance you can buy.

Those who want switch controlled outdoor lighting, should install all lights on one circuit. Use a 20 amp single pole switch, or install three way switches, Illus. 4, 98, one alongside your bed, another in the kitchen or front entry door. While darkness encourages, and light discourages an impulse to break in, if you can see what's going on, it makes for better shooting.

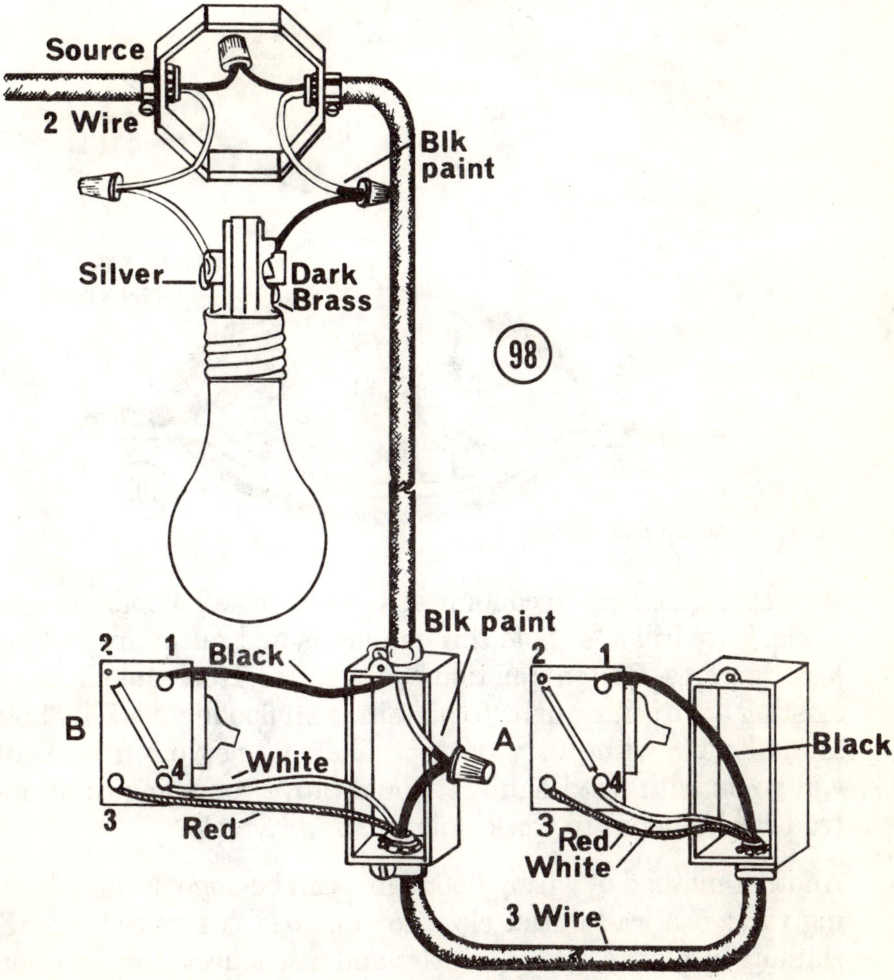

Another outdoor lighting installation that provides considerable peace of mind requires an extension cord, Illus. 99, containing a line switch. When a strange noise occurs in the middle of the night, you don't even have to get out of bed to turn on lights. To install a line switch, open cover and only cut the black line. Connect black leads to terminals, Illus. 100. Do not cut white line.

LINE SWITCH

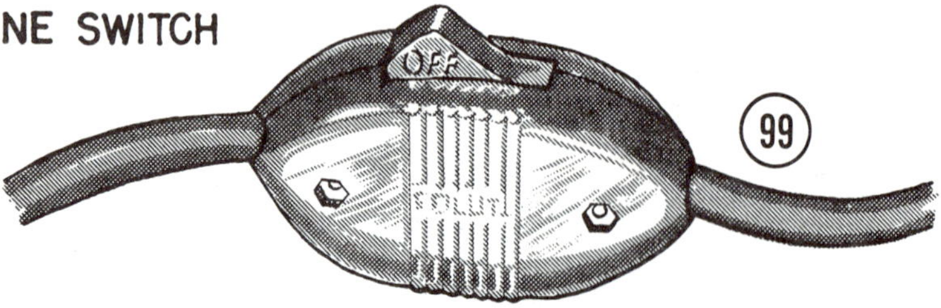

99

LINE SWITCH
(OPEN VIEW)

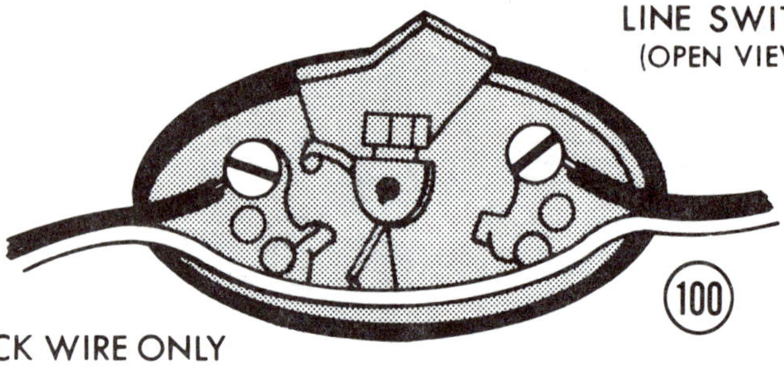

100

CUT BLACK WIRE ONLY

An easy, quick and economical way to install floodlights on gable, is to drill a ⅝″ hole through fascia and rafter, in position light requires. Fasten junction box to rafter. Run cable from an existing attic ceiling light to pigtails from floodlights. Plug hole around cable with putty or wood filler. While your attic light will go on with flood lights, it could prove confusing to an intruder who hopes to work quietly and privately.

Another method of wiring floodlights can be done in the following way: if a wall outlet close to your bed is serviced by BX running into attic, remove outlet and disconnect it in junction box. Use this cable as a switch loop, Illus. 24. Always remove fuse and trace circuit using a light and battery tester, Illus. 101. Or you can trace a circuit with a bell and battery, Illus. 102.

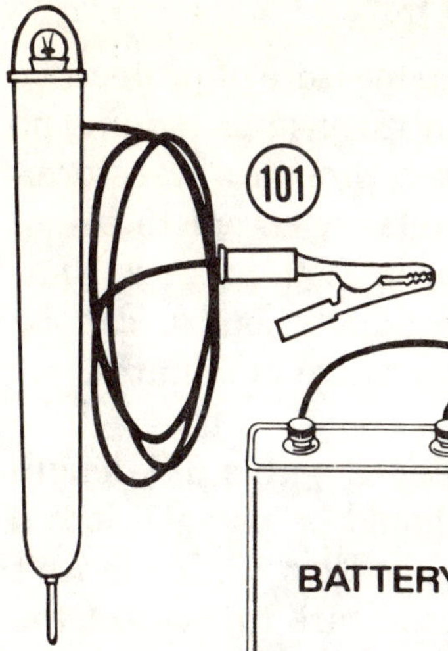

101

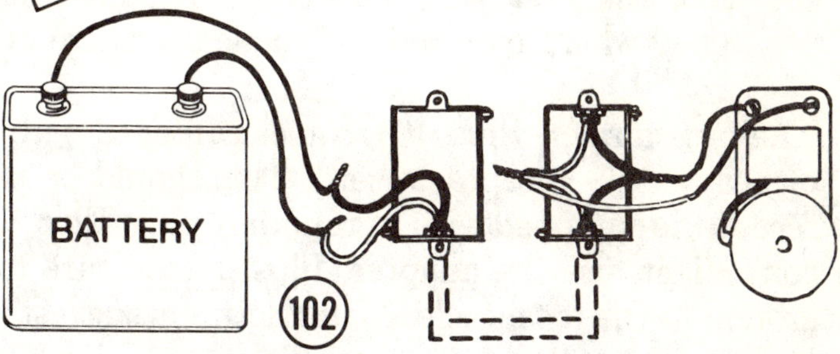

102

When converting existing cable to a switch loop, always make certain the power source passes through switch, then proceeds to light when switch is closed. Paint ends of white line black.

Another way to install outdoor lighting is by installing a weatherproof outlet, Illus. 75, outside a bedroom window. A wall switch installed in position shown, Illus. 103, provides an inexpensive way to control a lot of light. If you use an extension cord with a line switch, cover receptacle with a 1-Gang Telephone Wall Plate, Illus. 104.

103

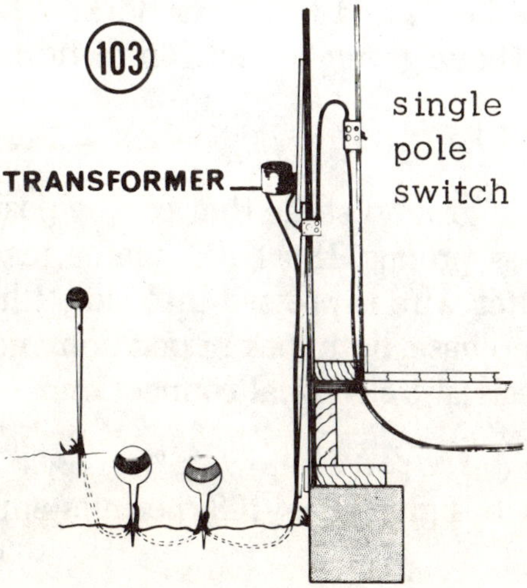

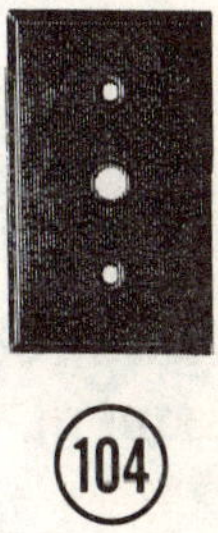

104

OUTDOOR DECORATIVE LIGHTING

Decorative lighting also provides a large measure of protection. Low voltage lighting systems are easy to install, require no special tools or experience. Manufacturers provide a transformer that converts 115 voltage to 12 volts. A complete set of extremely well designed plastic outdoor lighting fixtures, Illus. 103, 105, using 12 volt automobile headlight bulbs, can be positioned where required and made operative in minutes.

The transformer, Illus. 103, comes equipped with a 6 ft. length of heavy duty receptacle cord. This should be plugged into a receptacle equipped with a grounding plug, Illus. 51. If this isn't convenient, use the adapter, Illus. 17. Be sure to connect the ground to the receptacle. Manufacturers suggest mounting the transformer within 6 ft. of an outlet.

A mounting bracket, fastened to transformer, slips over two screws. These can be driven into side of house, or to a 2x4x4 stake, driven into ground. Install transformer at least 18″ above ground. In heavy snow areas, it should be installed well above level of snow.

Fifty feet of master wire, provided with each set of three lights, permits spacing lights where desired. Since outdoor lighting requires considerable planning, first lay master wire where you think it should go. Space lights and test after dark. While the master wire can be left above ground, we recommend burying same at least 3″ to 4″.

Each fixture comes with a ground stake that can be positioned where desired. Two brass prongs, Illus. 106, pierce insulation when the insulated master wire is pressed into slot, Illus. 107. By twisting stake to lamp base, both lock in position and make a permanent weather-resistant electrical connection.

An adjustment on back of light, Illus. 108, permits spreading light in area desired.

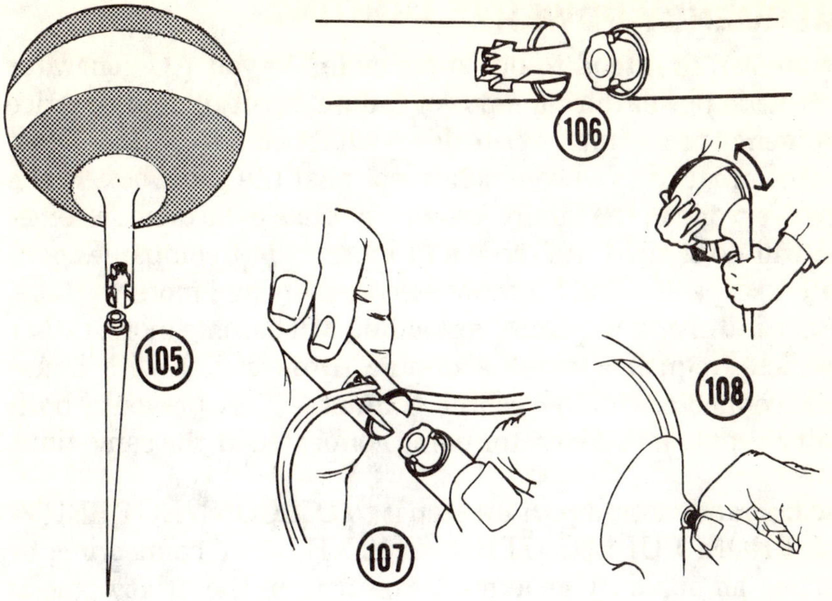

Since up to nine lights can be connected to a transformer, you can connect an additional 100 or 150 feet of master wire, and still power all lights from one transformer.

If a splice is required, strip insulation as shown in Illus. 109,* connect wire, and wrap wire in plastic tape, Illus. 110.*

109

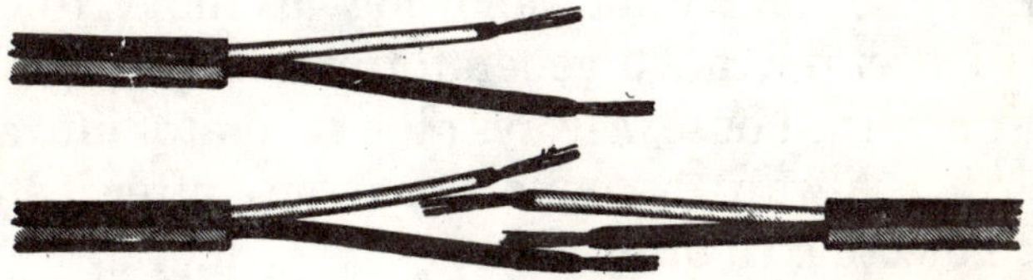

110

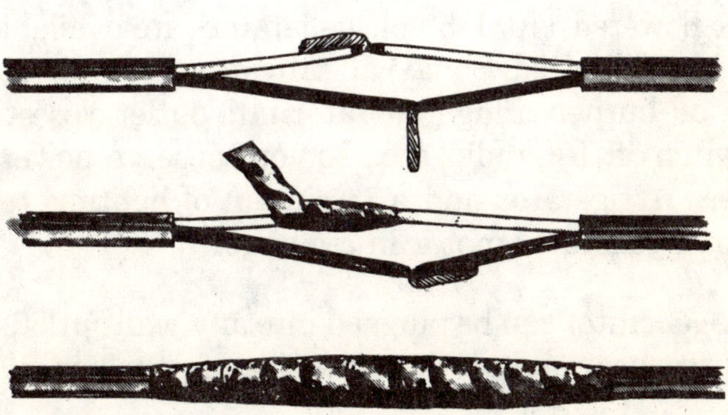

* Modifications required to conform to the Canadian Electrical Code - See pages 97, 98.

EMERGENCY POWER

If you own or intend to buy a portable 115 volt AC generator to provide power during a power failure, set it up and practice emergency procedures. Write down what needs to be done. Post this information in a prominent place near the panel box. Make sure everyone in the family knows what has to be done so someone will do it right. As more and more utility companies generate to capacity, and homeowners use more and more wattage, service interruptions from overloading will become more, rather than less frequent. Install a double throw switch with house load connected to the center contacts. This prevents both utility supply and generator being connected at the same time.

The first and most important step is to DISCONNECT SERVICE FROM PUBLIC UTILITY LINE before connecting, or starting an auxiliary generator. Power from the utility pole is disconnected by pulling handle to off position, Illus. 4; pulling main fuse block, Illus. 11, or snapping Main ON - OFF circuit breaker.*

A portable gas powered generator can provide sufficient power to prevent costly freeze-ups. Always plan ahead. Select a place in an open area, a detached garage, a toolhouse, or patio where generator can be run without allowing gas fumes to seep into the house. Never operate a generator in a basement or garage that's built into the house. Always plug generator into an outlet before starting. Always disconnect generator after it stops, and before reconnecting to utility line.

While car powered 110/115 volt generators are available, their capacity is still limited. Some will provide sufficient wattage to power an oil burner. Plug generator into outlet closest to fuse panel. Switch off TV, radio, etc., so you conserve power for the oil burner, refrigerator and a minimum of lighting required. Never run a generator motor in a garage.

While the generator can be plugged into any wall outlet, always use an outlet near and on the same side of panel, as oil burner circuit.

* Modifications required to conform to the Canadian Electrical Code - See pages 97, 98.

With main fuse disconnected, or handle on fuse panel in OFF position, with all lights and appliances, and particularly portable electric heaters disconnected, or switched OFF, an auxiliary generator will usually produce sufficient power to run an oil burner, a refrigerator and whatever lighting its power output will accommodate.

Auxiliary generators are rated according to wattage produced. Buy a size that can handle the equipment you want to use. Since wattage is lost in transmittal, don't overload.

ALWAYS REMEMBER TO: disconnect auxiliary generator before connecting to power line.

STRUCTURAL LIGHTING

The term structural lighting refers to any light source built into a home. It can be a lighted valance, cornice, wall bracket, cove, luminous ceiling, wall panel, or any of the other lighting ideas described. Because structural lighting complements furniture and furnishings, it has no style classification and cannot be dated. Its primary function is to lighten walls and ceilings to extend the visual area. Every installation can be made in two ways; by plugging into an existing wall outlet, or with concealed wiring.

Structural lighting contains a fluorescent channel, a continuous or multiple strip of channels installed in the various areas shown, Illus. 111 to 120. Each installation provides easy to build designer styled lighting.

CORNICE LIGHTING

This type of lighting is usually fastened to ceiling or to wall at ceiling height, Illus. 111. It can be installed over a window or on a blank wall. Since all light is directed down, it provides an ideal way to accent draperies, illuminate a picture wall, or provide a feeling of daylight when draperies are drawn at night. Complete details for building and installing cornices are outlined on page 71.

CORNICE (111)

VALANCE (112)

VALANCE LIGHTING

Customarily installed at window height, Illus. 112, these are built with a solid top to direct all light down, or with a diffused plastic top to allow light to bathe ceiling above. Valances are built to width of window, plus 12″ to 16″. This allows valance to extend 6″ to 8″ on each side, Illus. 113.

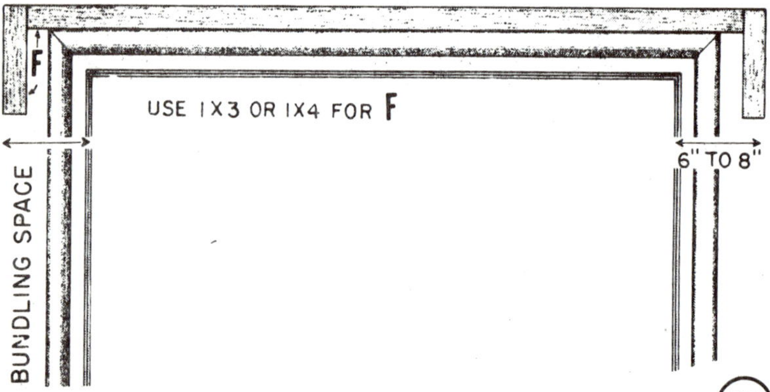

(113)

WALL BRACKET LIGHTING

These are similar to valances with one small exception. Wall brackets, Illus. 114, distribute light up, as well as down. When a valance is used over a large window, a wall bracket of equal length, installed on an opposite wall helps distribute and balance the illumination. When installed over a bed, a wall bracket is frequently installed 30″ above mattress.

When used to light a mural, or group of pictures, wall brackets are frequently installed 6″ above top row. Its length depends on its use. If used over a bed and installed parallel with its length, the bracket would be equal length.

LUMINOUS CEILING PANELS

This is one of the easiest and fastest ways to install a new ceiling, Illus. 115. Complete kits are available.

114 WALL BRACKET

115 LUMINOUS CEILING PANELS

LUMINOUS CEILINGS

This popular form of lighting has unlimited application, Illus. 116. A kitchen, bathroom, playroom, hall or foyer can be modernized at surprisingly low cost. Since little conventional ceiling board is required, the savings can help pay for plastic luminous panels.

116 LUMINOUS CEILING

117 LUMINOUS WALL

LUMINOUS WALLS

In new construction, a narrow hallway can be transformed into a path of light, small rooms can be made to appear larger by installing a wall of light, Illus. 117. Room dividers and privacy partitions can be built to screen a front entry door from living room, or provide privacy for a dining area. Refer to page 80.

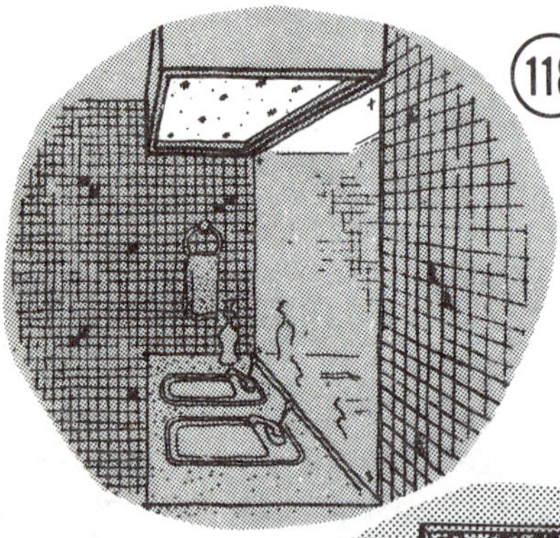

(118)

SOFFIT

COVE

(119)

(120)

SOFFIT, COVE, BOOKCASE LIGHTING

When light is required over a counter in a kitchen, bathroom or laundry room, use soffit lighting, Illus. 118. Refer to page 81.

Use a cove when you want to direct all light upward to ceiling, Illus. 119. This lighting is recommended for white, or near white ceilings. Cove lighting gives a soft, uniform lighting effect, but should only be used to supplement other lighting. Refer to page 81.

Bookcase lighting provides an excellent source of light when viewing TV, Illus. 120.

HOW TO INSTALL A WALL SWITCH AND FLUORESCENT FIXTURE

Fluorescent channel is one of the simplest forms of lighting to install. It consists of a baked white enamel fluorescent channel with white sockets or lamp holders at extreme ends. These come pre-wired in 18″, 24″, 36″, 48″ lengths. Since most channels are designed to butt end to end, they can be joined to provide a continuous strip of light. Do not space apart. To obtain proper distribution of light, fill cornice, valance, etc., with an equal length of fluorescent channel. Unfilled space between end of channel and end of cornice, up to 9″, creates no problem.

To install a wall switch and fluorescent fixture, Illus. 121, disconnect fuse serving circuit. Remove cover plate on outlet. Remove screws and pull outlet out of box. Using a screw driver, force top knockout out of box.

121

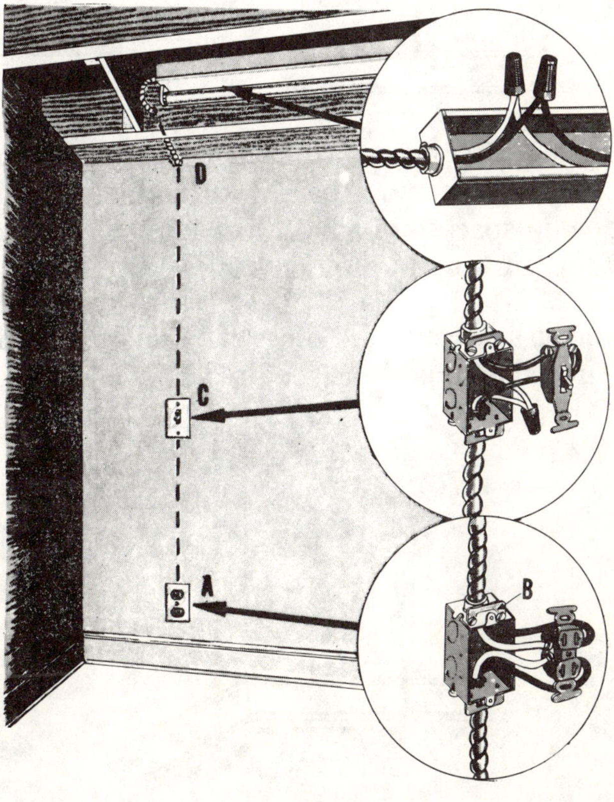

Measure distance from existing outlet to location selected for new switch. Carefully mark wall. Always measure distance from floor to an existing wall switch and place new switch at same height. Using a level, draw a guide line from existing outlet to point selected for new switch. Use the box template, Illus. 68, to draw outline of box. Saw opening where required.

If you are installing a luminous, or acoustic ceiling, and plan on installing fluorescent fixtures behind plastic diffusers, remove ceiling molding, and drill a hole where needed close to ceiling, Note D, Illus. 121.

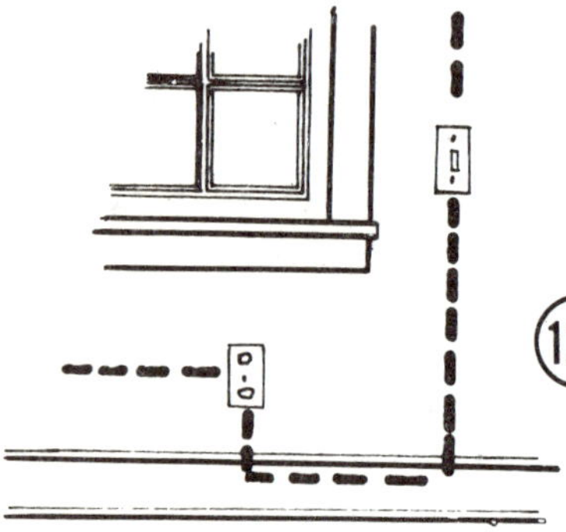

122

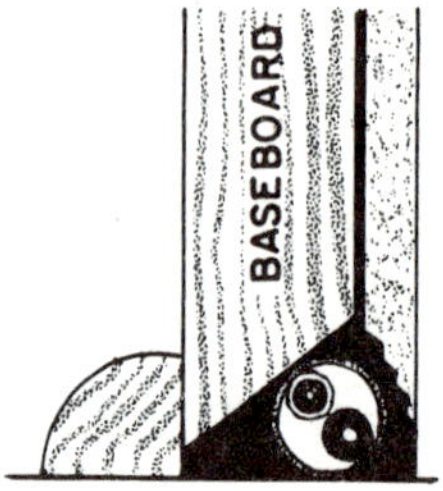

123

If no wall outlet exists in position shown, but there is one under a window, Illus. 122, remove baseboard and bevel plane back edge to provide space for cable, Illus. 123, or recess cable in plaster. Snake cable up in position shown.

If outlet is on an adjacent wall, Illus. 124, drill up from the basement and run cable to position required.

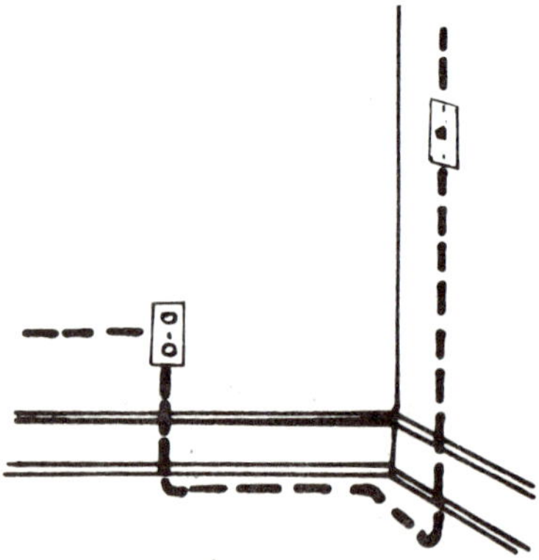

124

If you are installing fluorescent lighting behind a valance or cornice board, or in a luminous ceiling, measure distance from C to first fixture and add 12″, Illus. 121.

Fluorescent channel can be fastened to A in position shown, Illus. 125, with ⅝″ No. 7 roundhead screws.

125

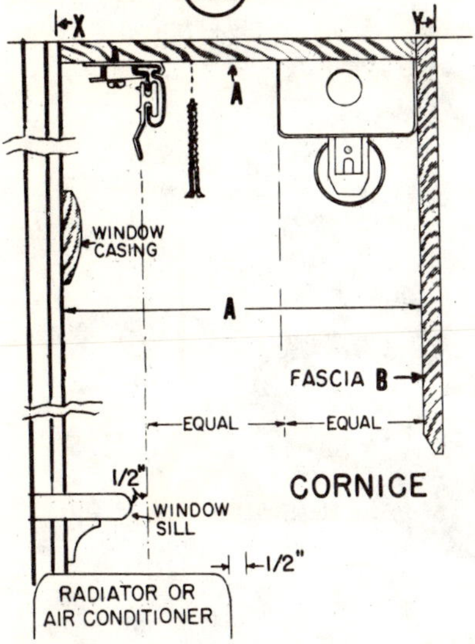

USE 1" LUMBER OR
1/2" OR 3/4" PLYWOOD FOR A

126

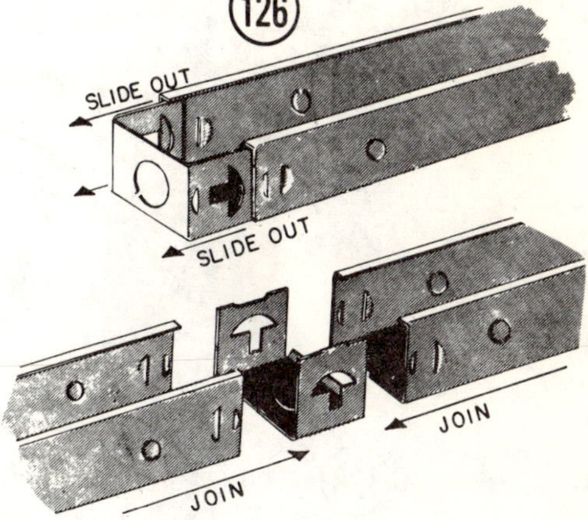

8 ft. channels are available. These accommodate two 4 ft. tubes.

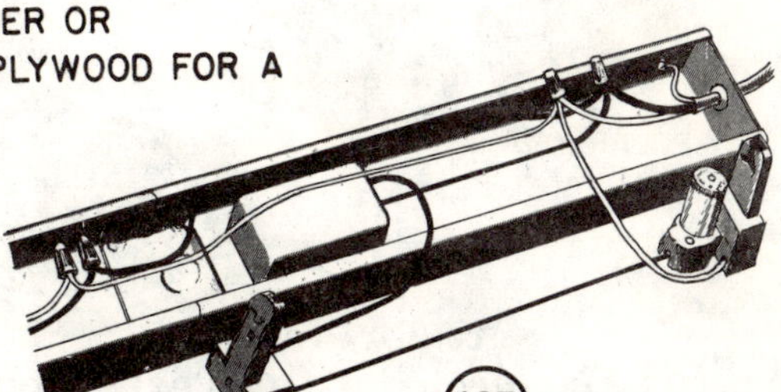

127

Fasten and wire channel end to end, Illus. 126, following channel manufacturer's directions, 127, 128.

If channel is to be operated with a plug-in cord and line switch, Illus. 129, connect necessary length of cord. Fluorescent lighting performs best when used on a grounded wiring system. Always use a grounding type extension cord to wall outlet.

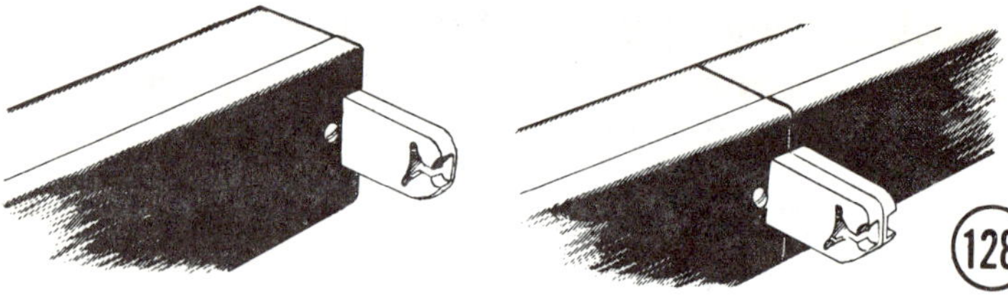

128

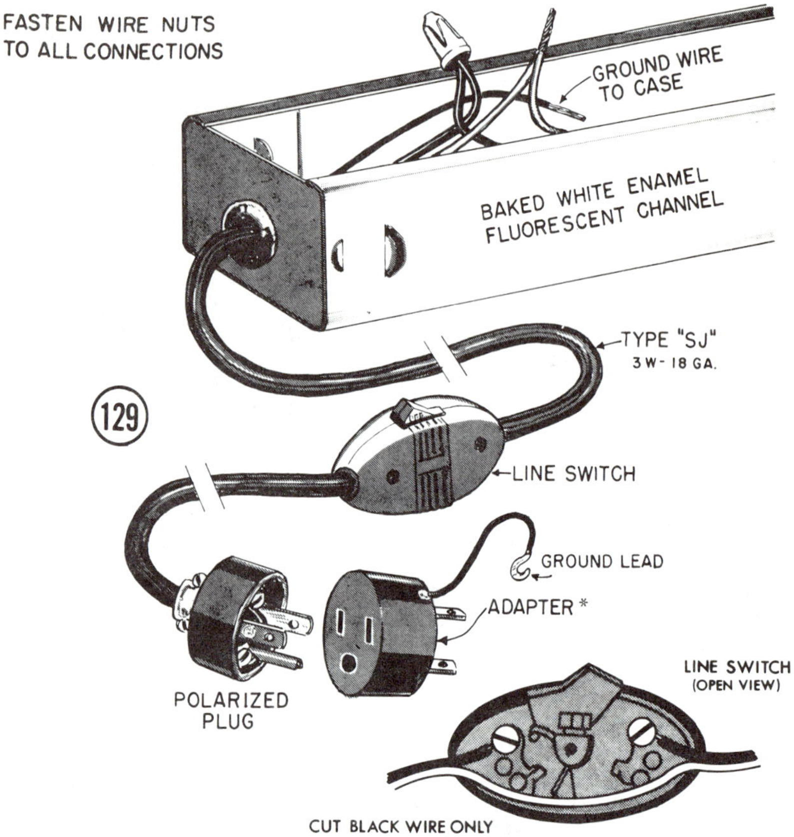

129

When fixture is to be connected with an extension cord, use 3w type "SJ" 18 gauge, heavy duty cord to connect channel to wall outlet. Use a polarized rubber plug and adapter. Install switch in line where it's most accessible. Insert adapter in wall outlet. Loosen screw on wall plate, insert and fasten ground lead to plate. Since the cord is usually placed behind draperies, it makes an easy and inexpensive installation.

Always remove tubes before fastening drapery track to A. When fastened in this position, draperies hang free of window

* Modifications required to conform to the Canadian Electrical Code- See pages 97, 98.

HOW TO BUILD AND INSTALL VALANCE AND CORNICE LIGHTING

Cornice lighting can be installed with or without draperies. It should be built full length of a room. Use length of fluorescent channel that fills cornice. Light from a cornice tends to lift the ceiling. It is especially helpful in rooms having less than an 8 ft. ceiling height.

Always install a cornice above a window when space between top of window frame and ceiling is less than 12″. Wall to wall cornices on opposite walls increase the illusion of height, make a room feel more spacious. Always remove ceiling molding in area selected before installing a cornice.

If one cornice in a room is used for lighting, while one or more in the same room contain both fluorescent channel and drapery track, follow this procedure.

Make A, Illus. 125, same width in all cornices.

To estimate width required, measure distance a window sill, radiator or air conditioner projects from wall and add ½″. Using a plumb bob, draw a line on ceiling at this point. This line indicates position of glides in drapery track. Measure width of channel. If it measures 2¾″, draw a line at 5½″.

To allow light to complement draperies, fluorescent channel should be placed approximately 2¾″ to 3″ from drapery track.

Draw a line on ceiling to indicate inside surface of fascia board. Cut A to width required. This now provides sufficient space to install drapery track beyond edge of window sill, and still provide necessary space for fluorescent channel.

If cornice is used only for lighting (no drapery track), 1x6 can be used for A. (When we specify 1x6, we assume it will measure approximately ¾″ x 5⅝″.

To obtain best results, install a cornice, wall to wall, and fill cornice with as much fluorescent channel as length permits. For example, if your cornice measures 10′1″, this permits installing two 4′ and one 2′ length with only ½″ unfilled space at each end. While three 3′ channels, with 6½″ unfilled space could also be used, the less unfilled space the better. Unfilled space up to 9″ at each end still provides adequate distribution of light.

(NOTE: Some fluorescent channels measure slightly longer than 2′, 3′ or 4′ in length.)

To build cornice, measure length required for A, Illus. 125. Cut lumber or plywood to size required. Since walls are seldom plumb, always measure space available at both X and Y. Cut A full length required, less 1½″. This allows for two ends C, Illus. 130. Always check end of board with a square before measuring for length.

130

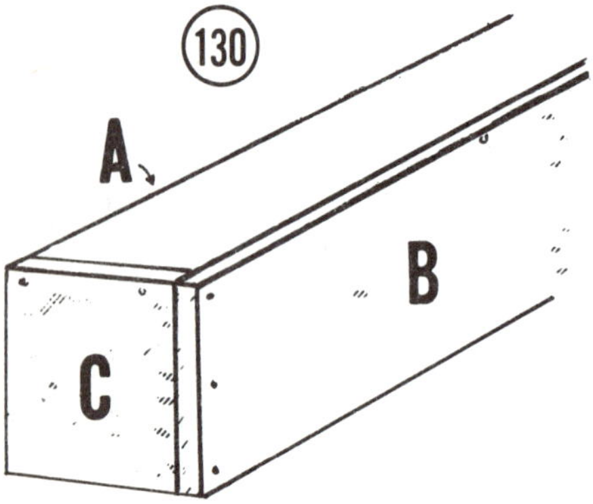

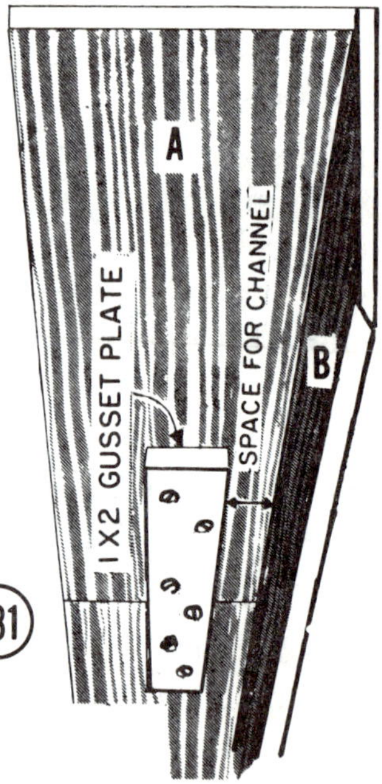

131

If two lengths are required, apply glue to ends, fasten in position with a 1x2 gusset plate, Illus. 131. Fasten gusset plate in position so it doesn't interfere with channel or drapery track.

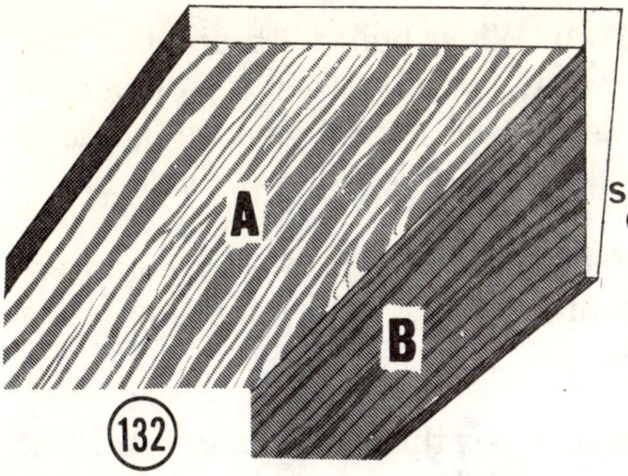

132

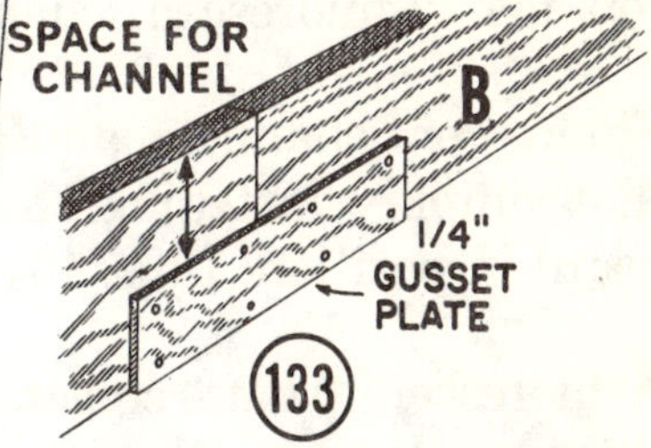

133

An 8″ fascia board (B), Illus. 131; 1x10, 1x12, or 8″ beveled cedar siding, Illus. 132, can also be used for long cornices.

When installed in a paneled room, cut fascia board from ¼″ plywood or hardboard. When lumber is used, plane bottom edge to shape shown, Illus. 125.

If cornice is to be painted, or covered with matching wallpaper, or fabric, lumber, ⅜″ or ½″ plywood can be used. If necessary to butt ends together to obtain length required, fasten end-to-end with a gusset plate of ¼″ plywood glued and screwed in position, Illus. 133.

Cut ends C to size required, Illus. 130. Plane bottom edge to shape of fascia board. Apply glue and nail C to A, B to A and C with 4 penny finishing nails spaced every 18″. Paint inside with two coats of inside flat white to provide necessary reflection of light. When dry, screw channel in position shown, Illus. 125.

Valance lighting is installed in the same manner as cornice lighting. While 8″ to 10″ valances are popular, many decorators now recommend fascias up to 12″. A valance should extend beyond edge of window frame a distance equal to one third of one draw curtain. If a draw curtain measured 24″, allow valance to extend 8″ for bundling when drawn back. Since the thickness of material is your best guide, estimate amount required by "bunching up" a curtain, then measure how much space it requires beyond edge of window frame.

Valances direct all light down. When built with an open top, they light up as well as down. If space between top of valance and ceiling is 10″ or less, use a solid top. If valance has solid top, install fluorescent channel in position shown, Illus. 125.

Fluorescent channel manufacturers now supply special hangers that simplify fastening channel to wall, and mounting fascia board in position, Illus. 134.

A fluorescent fixture consists of a box that holds ballast A, Illus. 135, lamp holders B, starter C and fluorescent lamp.

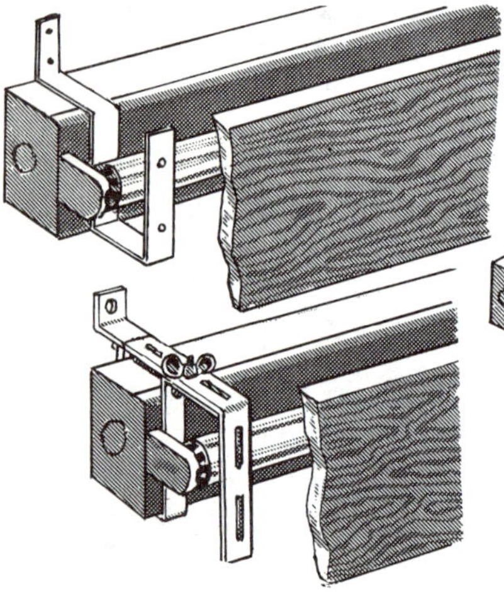
134

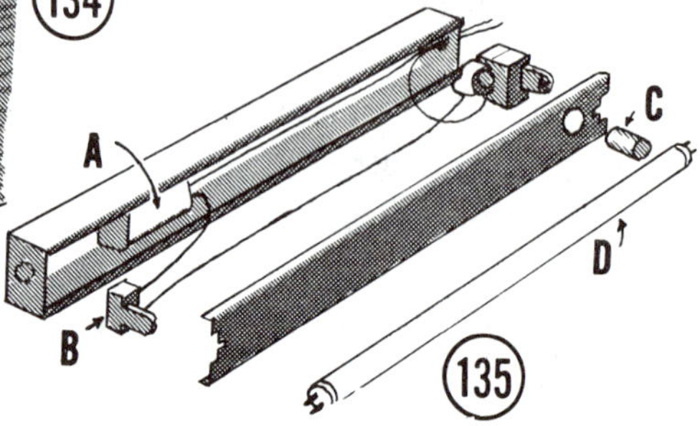

135

When purchasing channel, buy equipment wired with trigger, or rapid start ballast. These eliminate the need for separate starter.

Cover on channel is easy to remove. This permits fastening channel to ceiling or side of cornice. Pre-punched holes in channel permit fastening through either bottom or side.

Side mounted channel, Illus. 136, is also available. These are especially useful when you want to place lamp away from wall. This can also be done by using a 2x3, Illus. 137.

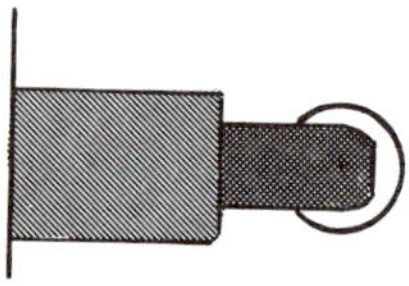
136

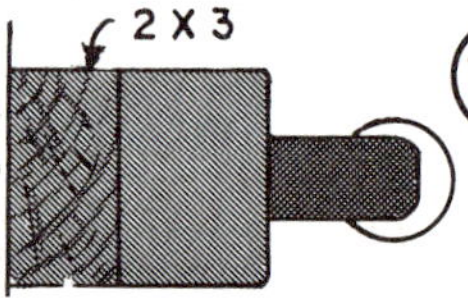

137

HOW TO INSTALL A LUMINOUS CEILING

Fluorescent channels can be installed over a smooth or badly cracked plaster ceiling, over plasterboard, etc., or between ceiling joists. While a luminous ceiling can be installed in any room, a finished height of not less than 7′6″ is recommended. In rooms where a luminous ceiling might project to 7′0″ or 7′3″, use a luminous panel, rather than cover entire ceiling.

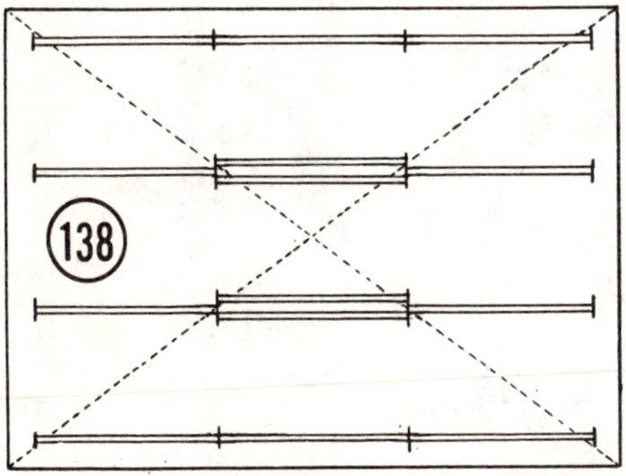

138

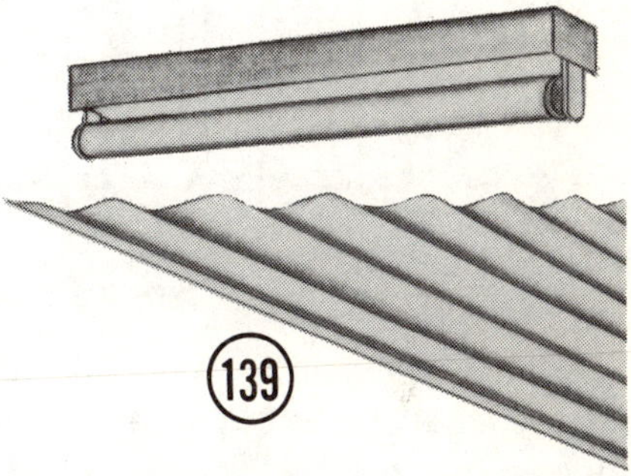

139

INSTALLATION OF LUMINOUS CEILING OVER PLASTER, PLASTERBOARD, ETC.

Remove lighting fixture on ceiling. Remove loose plaster. Paint everything you don't remove; all exposed pipes, heating ducts, conduits, etc., with two coats inside flat white. Measure area of ceiling, and draw a plan, Illus. 138. Draw diagonals to locate center of room, then space channels equal distance from center.

Space first row of fluorescent channel approximately 8″ from wall, space other rows 16″ or 24″. Channels can be run either across short measurement of room or long way.

If corrugated plastic panels are used as diffusers, Illus. 139, run channels at right angle to corrugation.

If you want more light over a mirror, counter, pingpong table, etc., install extra channels. Butt channels end to end to approximately 8″ from wall. Wire channels as shown, Illus. 140. Connect to switch used for overhead fixture, or install wall switch.

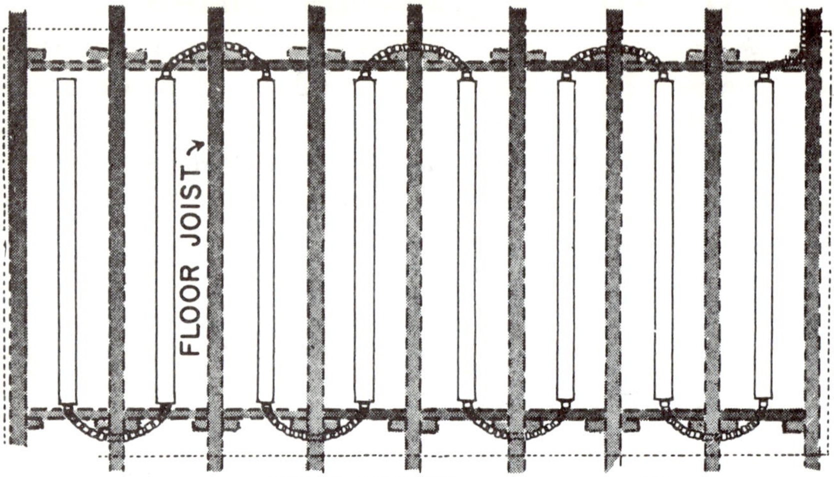

140

If you install extra channel to provide more light for special occasions, control these channels from a separate switch.

Companies producing luminous ceiling panels provide the aluminum framing required. This consists of an angle, or L extrusion, T-rail, Illus. 141, plus other interlocking members that fasten to wall and ceiling at height selected for new ceiling.

With level, Illus. 142, draw a level line around room at height selected for new ceiling.

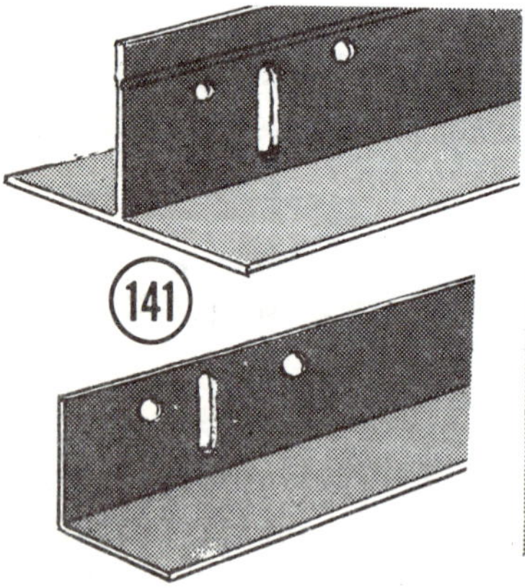

141

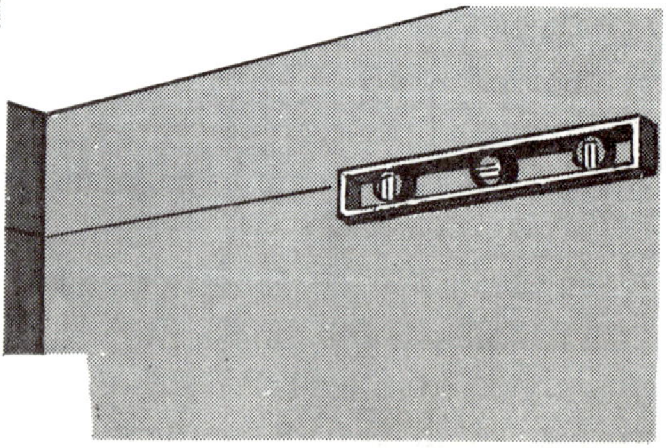

142

143

Measure 16″ from a wall to locate stud in wall. Test with a 4 penny nail. When you locate a stud, measure every 16″ to locate other studs. Mark wall to insure fastening angle to studs with nails or screws supplied by manufacturer, Illus. 143.

Other parts provided for luminous ceiling lock together to form 2x2 squares, or 2x4 rectangles. Cross members, or T-rails, are supported by hangers that fasten to joists, Illus. 144. The panels are then dropped into position.

144

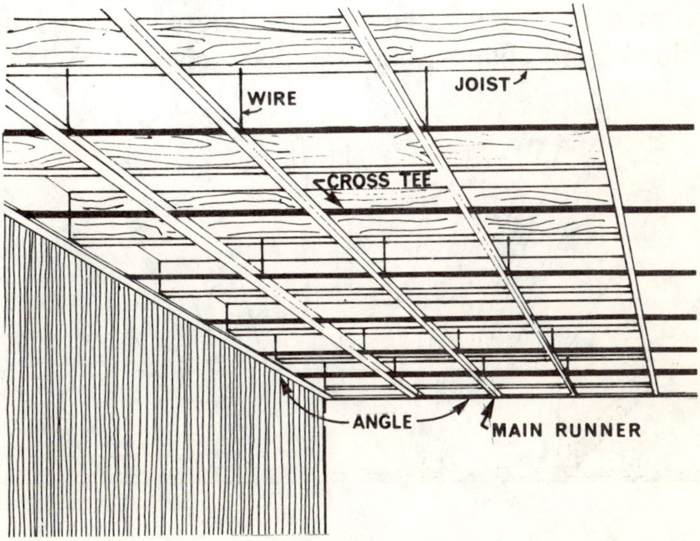

LUMINOUS CEILING PANEL

Illus. 115 shows a luminous ceiling panel installed in a playroom measuring approximately ten by twenty feet. Since ceiling joists measured under 7′6″, a panel was installed rather than a luminous ceiling.

Measure room and draw a plan. Locate center, space and nail 1x2 furring across ceiling joists. Install acoustical ceiling tiles around perimeter of ceiling and in area not covered by luminous panel, Illus. 145. A 5′6″ x 11′0″ luminous panel was installed in area shown. It accommodated eight 4′ channels.

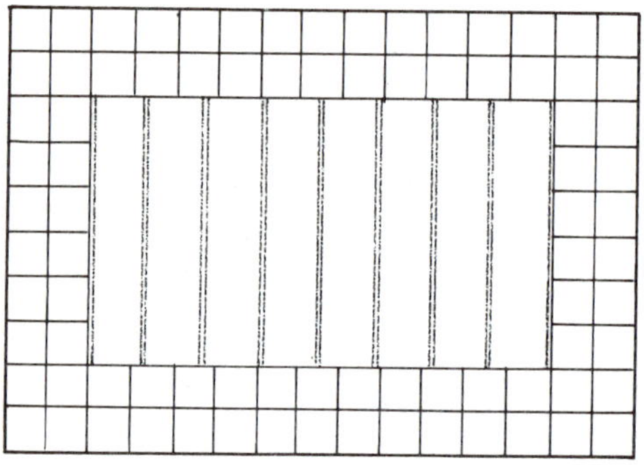

145

When modernizing a basement with limited ceiling height, install fluorescent channel between floor joists, Illus. 146.

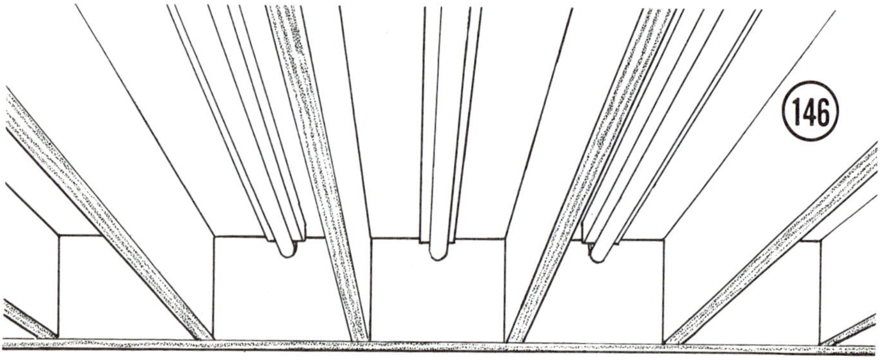

146

After deciding length of channel, number, and location, the next step is to box in ends of "light cavity" with ¾" plywood, or lumber cut to size required to fit flush with bottom edge of floor joists, Illus. 147.

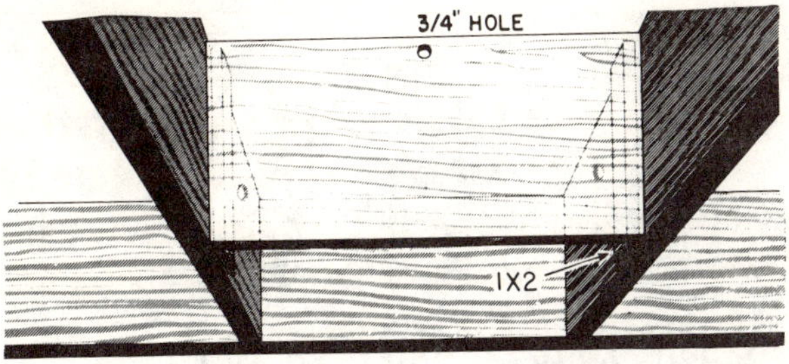

147

148

With a square, draw lines on joist to indicate length of channel, add 3", (51" for four foot channel). Drill ¾" holes in position noted in ends. Drill ¾" holes through joists, 3" up from bottom to permit snaking cable.

If 1x2 or 5/4 x 3 (furring) is nailed across bottom of joists to provide a nailing grid for acoustic ceiling tiles, cable can be stapled to bottom of floor joists between furring.

Note wiring layout, Illus. 140. Where codes require, cut sheet asbestos to size required to cover ceiling, sides, and ends of light cavity. Glue or staple in position. Or asbestos cement board can be sawed, drilled, and nailed in position. If asbestos board is used, paint flat white. Also paint any pipes, conduit, etc., running through light cavity. Fasten fluorescent fixture in position with ¾" No. 7 roundhead wood screws.

If bridging between floor joists, Illus. 148, prevents installing channel within area selected, move, and fasten bridging outside light cavity.

If floor joists are 2x8, fasten channel in position , Illus. 149.

If joists are 2x10 or 2x12, mount channel on a 2x3 or to 1x2 cross braces, Illus. 150.

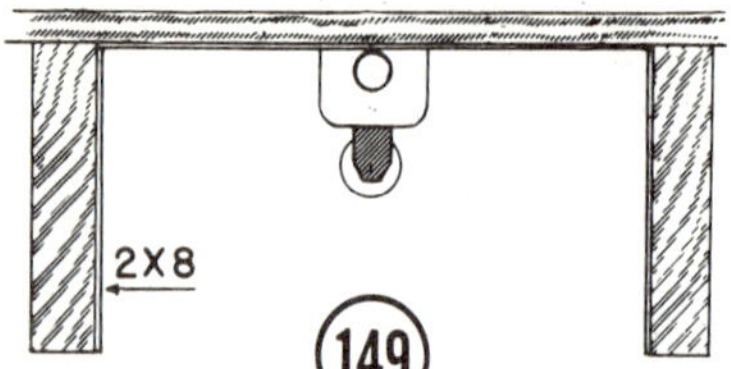

149

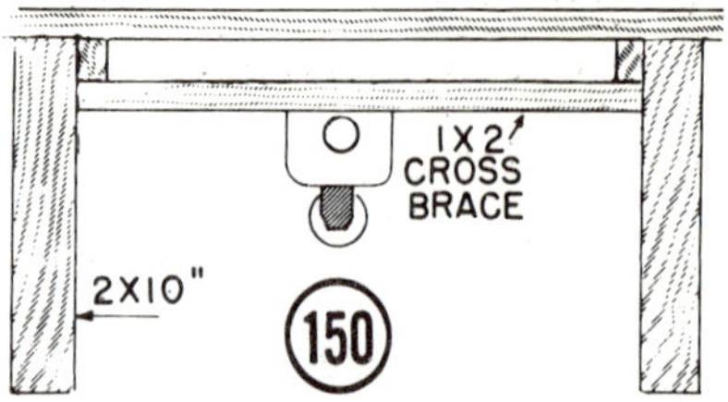

150

If floor joists are 2x6, nail 2x2 to bottom edge, Illus. 151.

Always install fluorescent tube approximately 3″ to 5″ away from plastic diffuser, Illus. 152.

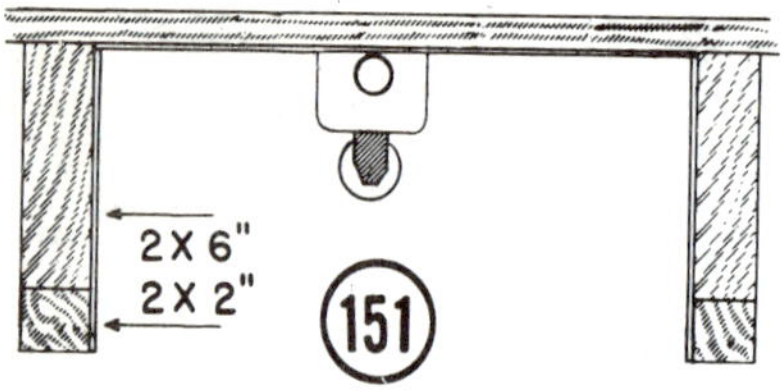

151

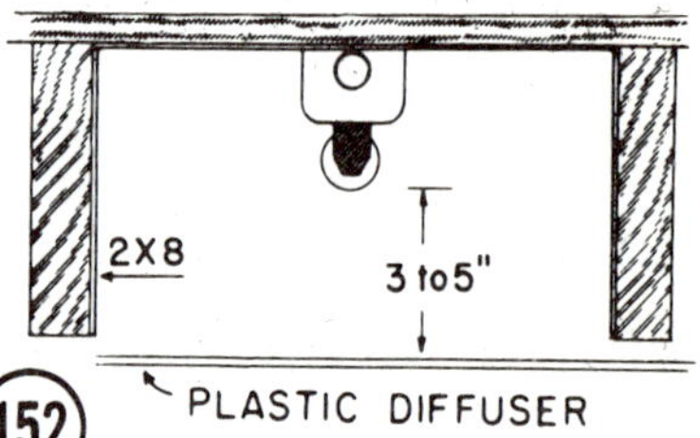

152

LUMINOUS WALL

A luminous wall increases the feeling of spaciousness. Illus. 153 shows installation of channel on 2x4 studs. Decorative plastic panels are held in place with ⅝″ x 1⅝″ round edge casing in position shown with 1″ No. 6 roundhead screws. A privacy partition or room divider can be built using 2x6's, 16″ on centers as shown in Illus. 154. This is an ideal installation for walls between hall and living room.

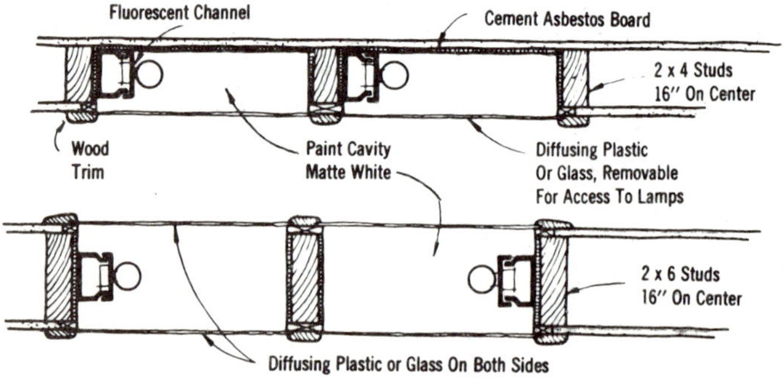

153

154

SOFFIT LIGHTING

A soffit, Illus. 155, 156, 157, can be built 16″ to 24″ in width. Use two rows of channel for soffits up to 18″ in width; three rows for 24″. Cut ½″, ⅝″ or ¾″ plywood to width and length required for top. Cut 1x8 or width fascia board desired to length required. Cut ends to size required. Nail 1x2 to top in position shown. Glue and nail ½″ x ¾″ shoe molding to bottom edge of fascia and ends, Illus. 157. Nail fascia and ends to top and to 1x2 with four penny finishing nails. Apply sheet asbestos, install channels, test wiring following directions previously outlined.

Nail clam shell casing to studs in wall in position noted, Illus. 156. Cut plastic diffuser to size required. If you plan on installing an "L" shaped soffit, cut plywood top 16″ or wider by length required to permit screwing top to joists.

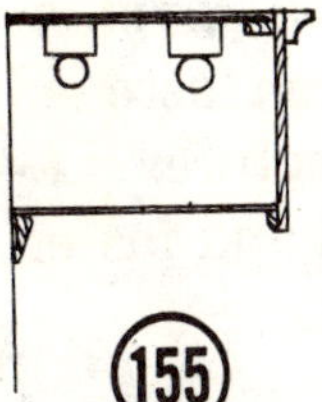

155

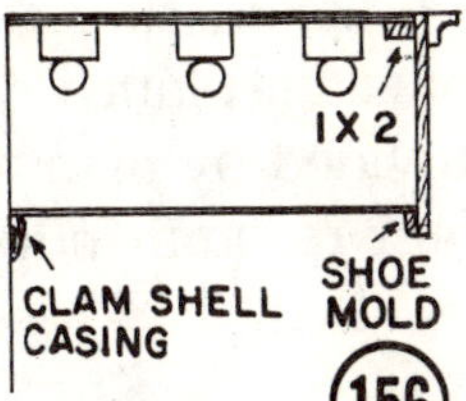

156

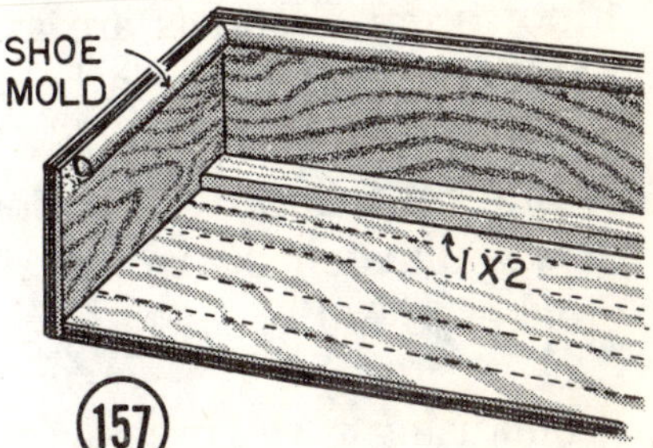

157

COVE LIGHTING

An easy way to install cove lighting is shown in Illus. 158. A 2x3 is nailed to studs in wall 12″ or more down from ceiling with 16 penny common nails. Place cove as far from ceiling as possible to allow even distribution of light. Fasten channel in position.

DRILL 2X3 FOR 16 PENNY NAILS

158

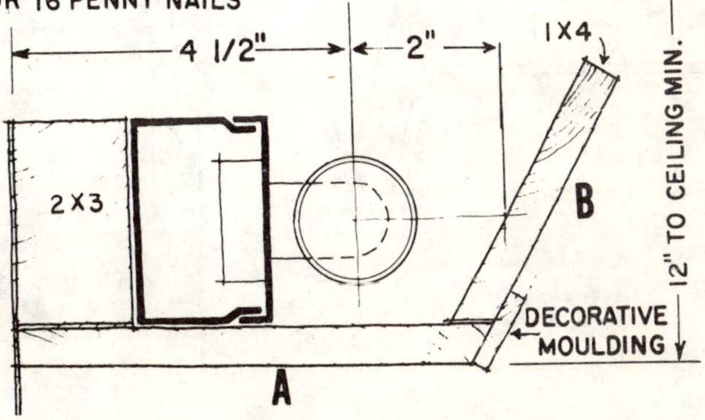

Cut a 1x8 for bottom A to width and angle shown. Cut 1x4 to width and angle shown for B. Glue and nail A to B. Glue and nail decorative molding in position shown. Paint inside flat white. Nail A to 2x3.

159

LIGHTED WALL BRACKET

Fluorescent channels are available with special brackets that permit mounting fascia board to bracket, Illus. 134. A 1x6 or 1x8 is cut to length required, bottom edge planed, then fastened to bracket following directions provided by manufacturer. Locate and fasten 1x4 to wall at height recommended and fasten bracket to 1x4.

When using a wall bracket over a bed, in a corner, or over twin beds, Illus. 159, build bracket to complementary length. Locate fluorescent tube 2″ from bottom edge of fascia board. When installed over a bed, the bottom edge of fascia should be 30″ from mattress.

160

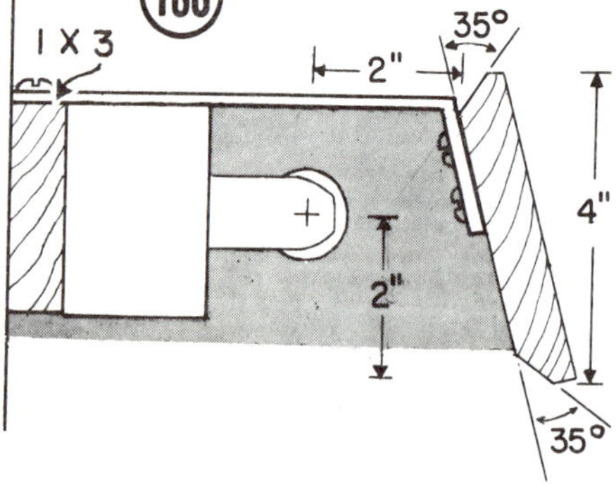

161

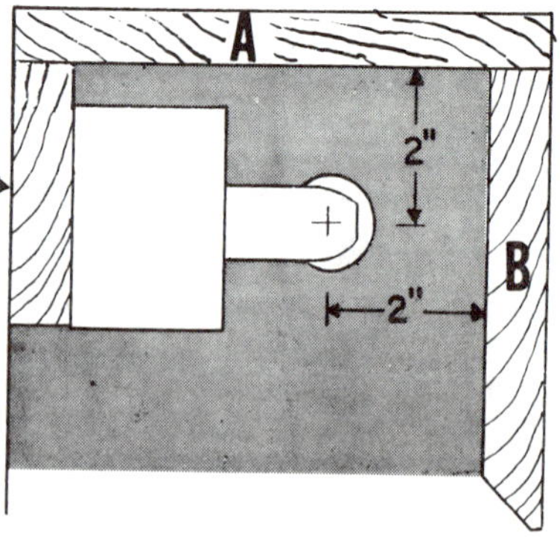

A slanted wall bracket, Illus. 160, can be made and installed in the following manner. First estimate length of wall bracket required. If used over a bed, make wall bracket same width as bed. Cut 1x3 to length equal to fluorescent channel. Paint exposed edges white. When dry, nail in position to studs with 8 penny common nails. One nail to each stud should be sufficient. Fasten channel to 1x3. Install line cord and switch.

Follow procedure outlined for cornice if you want to make a lighted wall bracket that directs all light down. Top A need only be wide enough to place fascia 2″ away from center of tube, Illus. 161. A slanted wall bracket, Illus. 160, does a better job of spreading light for reading.

162

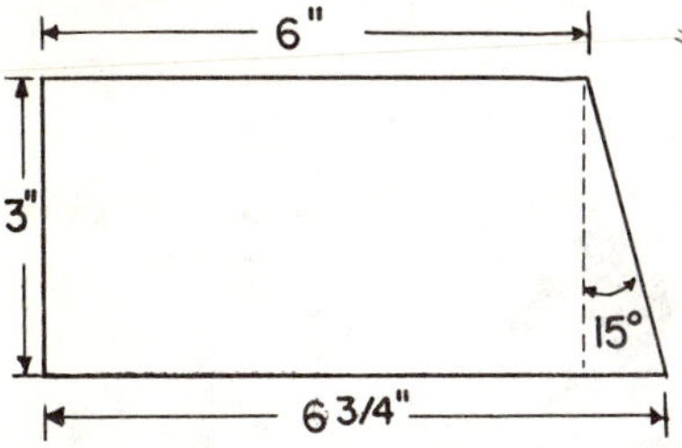

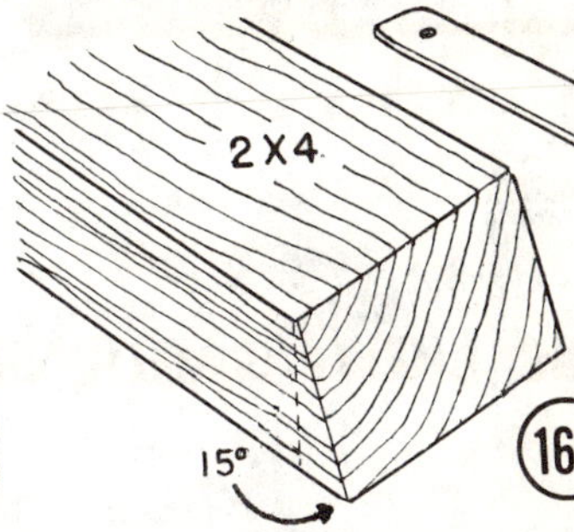

163

164

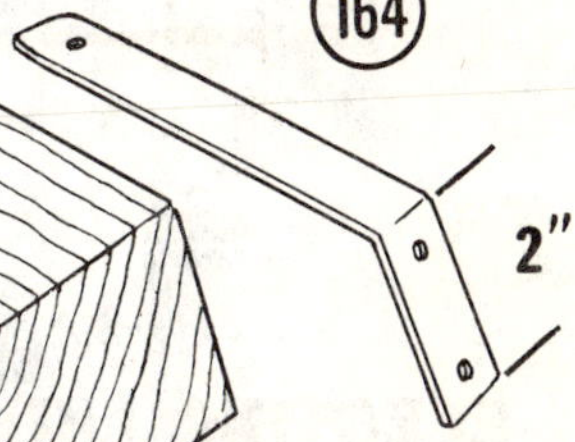

Cut fascia to length of channel plus 1½″. Cut ends to shape shown, Illus. 162. Plane bottom edge of fascia B to angle shown. If you want tube in wall bracket to project further from wall, use a side mounted channel, Illus. 136, or a channel mounted on a 2x3, Illus. 137.

You can make brackets by cutting ⅛″ x ¾″ aluminum or steel bar stock 8″, or length required. Drill 5/32″ holes where indicated for ¾″ No. 7 roundhead screws. Make form by cutting one end of a 2x3 or 2x4 to 15° angle, Illus. 163.

Bend 2″ of bar to angle, Illus. 164. Make three brackets for a 4 ft. channel, five brackets for an 8 ft. channel. Paint inside face of fascia and ends flat white. Fasten bracket to fascia at ends and center with ¾″ No. 7 roundhead screws. Bracket can also be fastened to channel with ⅜″ self-tapping screws. Fasten brackets to 1x3 with ¾″ No. 7 screws.

A wall bracket under cabinet provides great convenience. Drill and fasten length of 1x1 to bottom of cabinet in position shown, Illus. 165. Screw or nail 1x3 to studs. Fasten channel to 1x3 in position shown. Cut fascia from ¼″ plywood or 1″ lumber to width shown and to length required. Paint 1x1 and inside face of fascia flat white, when dry, glue and screw fascia to 1x1 with 1¼″ No. 7 flathead wood screws.

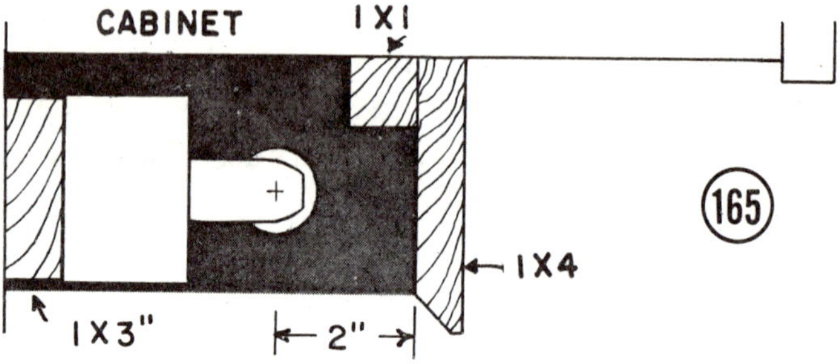

165

ELECTRIC LIGHT GARDENING *

Apartment dwellers, homeowners and almost everyone with a few spare feet of space, can enjoy year round gardening, thanks to photosynthesis, or as the Greeks called it, "light." Electric light permits a lot of fun gardening wherever there's a wall outlet.

*From facts provided by General Electric Lamp Business Division.

Much has been accomplished in this field by the U.S. Department of Agriculture's, Plant Industry Station, and State University Research Centers. Results from these studies now simplify growing big, healthy, beautiful plants in basements, garages and apartments.

Using fluorescent and incandescent lamps, experimental stations turned up some amazing information. Chrysanthemums delayed blooming when subjected to longer hours of light, natural or electric, while tuberous begonias bloomed earlier when subjected to longer hours of light. Food plants grew larger, faster, when bathed in light.

The daily duration of darkness was also found to be a controlling factor in growth and bloom. By interrupting periods of darkness, with only limited periods of light, blooming was materially affected. Results from these experiments now enable commercial growers to schedule blooming of selected popular plants at stated intervals, to meet the demands of the market.

A case in point is poinsettias and chrysanthemums. Each can be brought to bloom at the very time demand is greatest. Several crops of certain plants can be grown in one year instead of two.

Field tests indicate a mixture of fluorescent and filament provides better results than fluorescent lamps alone. A good ratio is 10% of filament to 90% fluorescent. While research has produced much information, amateurs are still making many interesting discoveries and are witnessing fascinating results.

When lighting an area from a short distance, with a single fixture, you increase the intensity four times, when you move the plant halfway from the lamp.

A single fluorescent reflector fixture containing two 40 watt lamps provides from 260 to 860 foot candles at 6″, Illus. 166, 330 to 500 at 12″; 350 to 380 at 18″. When reading above chart note 3″ spacing from a line drawn down center of fixture.

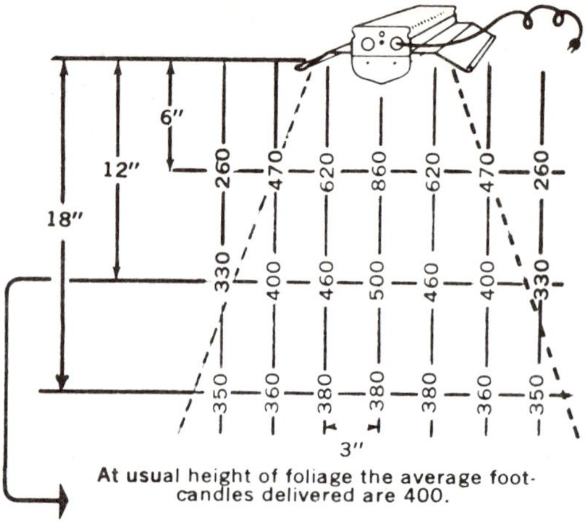

(166)

Filament bulb reflectors, Illus. 167, 168, should be spaced from one to one and a half times their distance from all plants. This insures uniform growth.

(167) **REFLECTING EQUIPMENT**

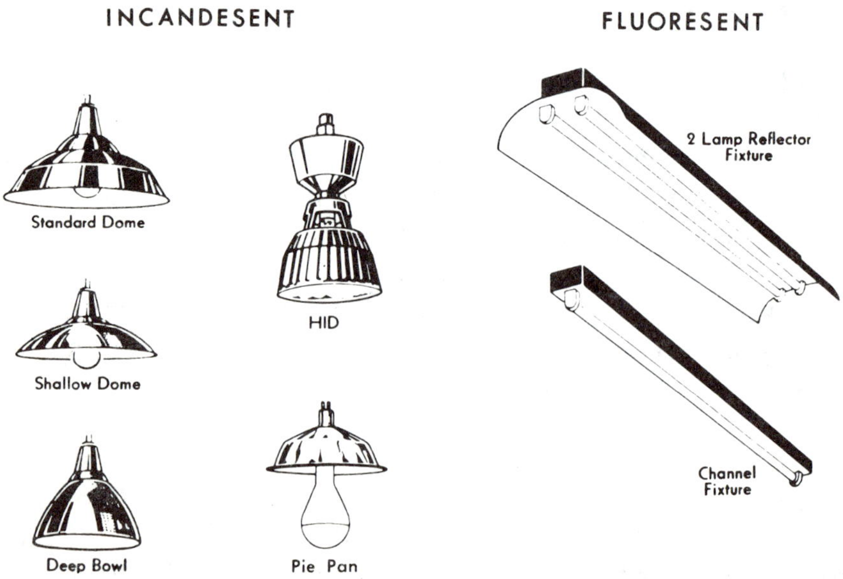

LAMP SIZE AND LOCATION OF REFLECTORS TO OBTAIN 10 FOOTCANDLES ON BENCHES
(Filament-lamp system)

Lamp Size Watts	Mounting Height— Bottom of Reflector Above Bench	Spacing	Area Square Feet
25	2′	3′ x 3′	9
40	2′ 8″	4′ x 4′	16
60	3′ 4″	5′ x 5′	25
75	4′	6′ x 6′	36
100	4′ 8″	7′ x 7′	49
150	6′	9′ x 9′	81
200	6′ 8″	10′ x 10′	100
300	10′	15′ x 15′	225
500	13′ 4″	20′ x 20′	400

168

Correct spacing of filament bulbs is important, Illus. 169.

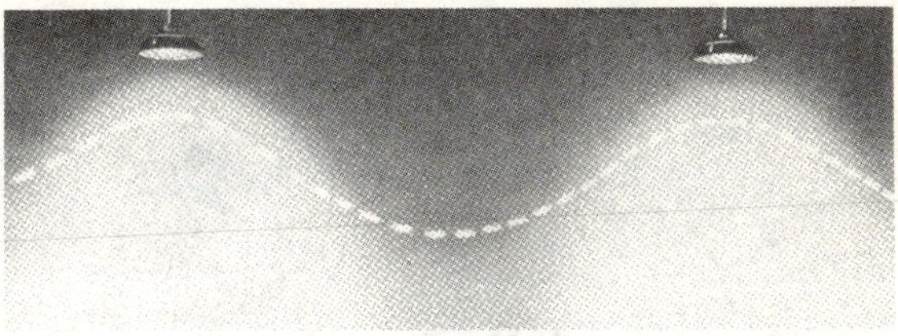

Too wide spacing produces non-uniform illumination with low spots between fixtures.

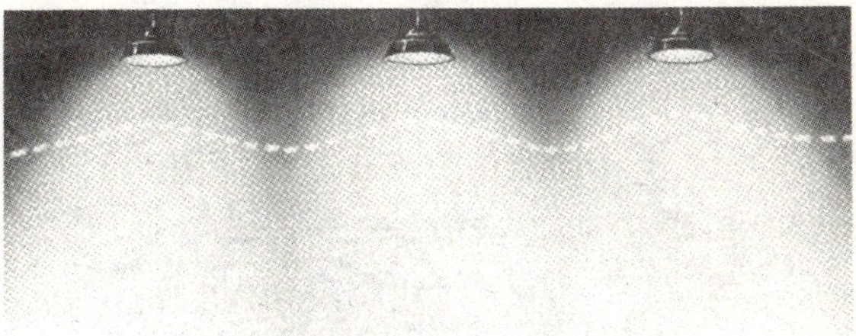

169

Correct spacing eliminates such relatively dark areas.

Illus. 170 shows a 26″ wide greenhouse with 100 watt lamps place 5′0″ above plantstand.

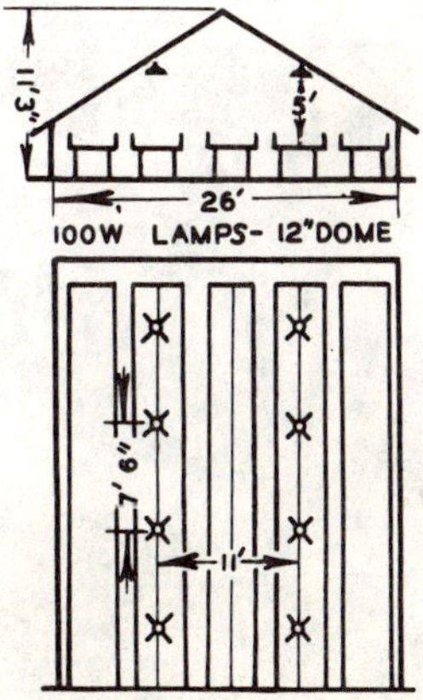

170

A footcandle meter, Illus. 171, registers amount of light.

171

1″ slimline daylight or cool white fluorescent lamps, Illus. 172, placed in position above plants as noted in Illus. 166, makes this space saving cellar installation a real producer. Cuttings and seedlings can be grown the year round with no daylight.

172

African violets, grown under light timer controlled lighting, Illus. 173, were rated tops. Tests conducted by the Ohio State University, in a windowless basement, in which a temperature of 65° and humidity of 60% were maintained, produced results indicated, Illus. 174.

173

174

	LIGHT	600 FOOTCANDLES		
	TIME	6 HRS.	12 HRS.	18 HRS.
AVERAGE NUMBER PER PLANT	LEAVES	44.6	54.3	55.7
	FLOWER STALKS	18.9	22.6	28.3
	FLOWERS	92	180.8	239

While plants differ in their basic needs, some grow better in bright sunshine, others do better in the shade, all require natural or electric light to transform the various elements into living organic matter. Plant derive their basic source of energy from light, as well as from the soil, moisture, fertilizers, temperature, air and humidity.

Through a combination of electric lamps, unbelievable results in producing all of the sun's rays – visible, ultraviolet and infrared, was achieved in research by the electric industry. Tests indicate that visible light is one segment of the spectrum that offers major benefits. Tungsten filament lamps, Illus. 175, and fluorescents are the most popular. The simple reflectors, Illus. 167, are used with filament bulbs, while a single or double tube reflector can be used with fluorescent bulbs. Filament bulbs ranging from 60 watt to 500 watt are recommended.

PS 30 lamps, Illus. 175, provide a wide-flood distribution that uniformly covers four times as much area as a standard 100 watt lamp-reflector combination. Both the R and PS 30 lamps should be protected against dripping or splashing, or from rain and snow when used outdoors.

175

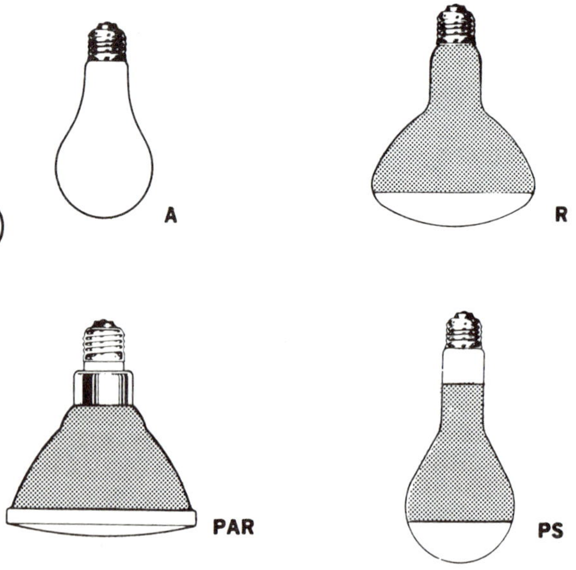

PAR lamps are made of thick glass. These can be used out of doors.

Use a combination of fluorescent and filament light when growing plants indoors. One of the most intriguing facets of "light gardening" is the results achieved from various colored bulbs. While the daylight and standard cool white lamps produce fairly uniform results, and are recommended for all beginners, plants respond to colored light. Blue lamps, when used alone, cause low stocky growth. The plants may be well-filled out, but tend to grow squatty. Red light tends to create a tall and spindly plant. A proper balance of red and blue rays produces plants that have normal growth and shape. Green light has no important photoperiodic effect on most plants.

The question of how much light a plant needs varies with the plant. Daylight out of doors may register as high at 10,000 foot candles according to a bulletin released by General Electric. Natural light in a home during the winter may vary from 10 to 1000 foot candles. Many varieties of plants have been grown to maturity in artificially lighted growth chambers at from 300 to 2500 foot candles.

African violets grow best in light registering between 600 to 1000 foot candles. A few hundred more foot candles provided even better than average results. Ohio State University has done a considerable amount of research on African Violets and offers some interesting bulletins on the subject.

Two 40 watt fluorescent lamps, under a reflector, Illus. 166, will produce 860 foot candles of light within 6″; 500 foot candles within one foot. This is considered satisfactory for seedlings. Since humidity is an important factor to plant growth, the amount of moisture in the air and soil should be watched. Your garden supply and seed retailer can advise you regarding the amount of moisture required.

Two 40 watt fluorescent lamps will usually provide sufficient light to cover a planter 3′0″ x 6′0″.

WINTERIZING YOUR ROOF

To prevent snow and ice from damaging your roof and gutters, install roof de-icing tape lines, Illus. 176. These are available in various lengths from 20 ft. to 100 ft. One tape should be lowered down the full length of each leader, preferably below ground. The balance should be laid full length in each gutter.

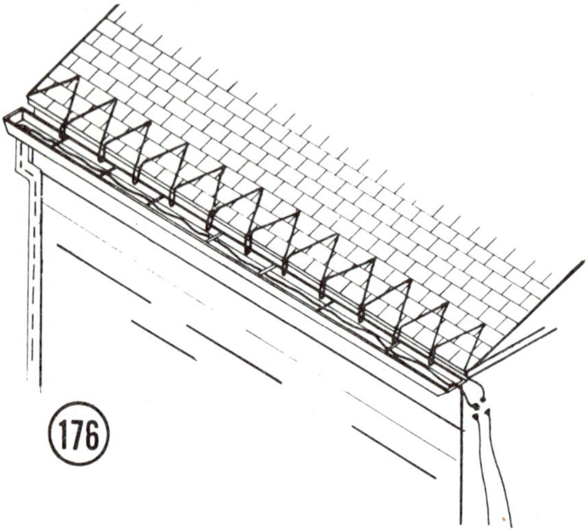

176

Another tape should be laid out and fastened to roof following manufacturer's recommendations, Illus. 176. To obtain adequate protection, fasten tape to roof every 2 ft. in position shown along the third shingle course from eaves.

The manufacturer of heating tapes provides sharp clips, Illus. 177, that are slipped under asphalt shingles. We don't recommend these for wood shingles. Brass screw hooks, Illus. 178, can be used. Screw in position and daub cement over base. The hooks permit removing tapes in the spring.

177

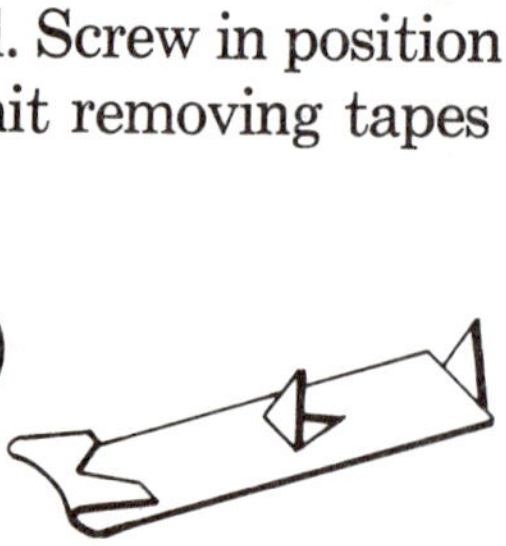

178

Always buy heating cord to full length roof, gutter and leader require. Lay it out according to manufacturer's directions.

Never allow a heating tape to overlap. Plug heating tapes into a waterproof outlet. Check the wattage of tapes. Carefully check the circuit they are plugged into. Estimate wattage normally used and don't overload the circuit. Never connect any appliance or tape that causes an overload. Costly fires can result. Be sure to use a heavy duty extension cord. A No. 12 extension cord can be plugged into a 20 amp fused circuit. Most inexpensive extension cords, even those advertised as "heavy duty" should not be used where a long length of tape is required.

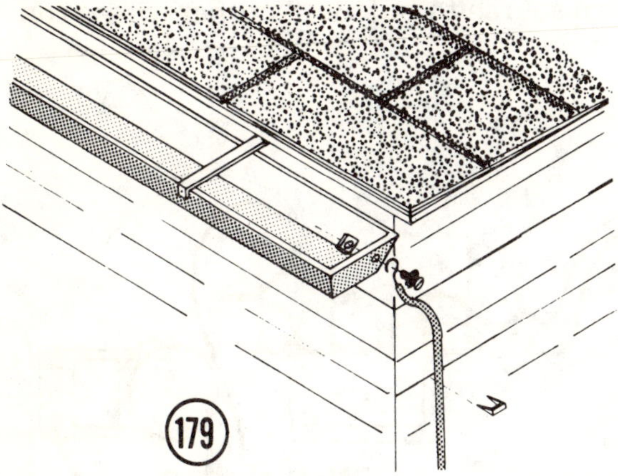

179

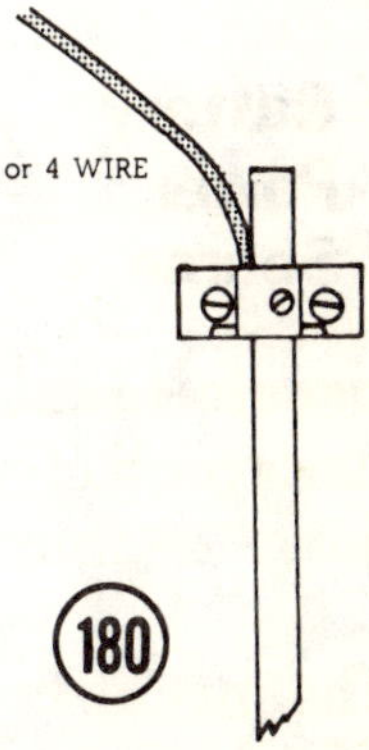

180

Always buy strings of lights that have the UL (Underwriters' Seal of Approval)*.

While most gutters are grounded by leaders running into the ground, play safe. Drill a hole in gutter, Illus. 179, and fasten a No. 6 or No. 4 gauge solid conductor insulated wire to a pipe in the ground, Illus. 180.

Always remove heating tapes in the spring as the hot sun won't do them a bit of good.

*Modifications required to conform to the Canadian Electrical Code- See pages 97, 98.

DOOR BELL INSTALLATION

Most builders install a transformer, Illus. 181, to power a door bell or chime. This plugs into a 115 volt outlet. Door bell wiring is connected at B.

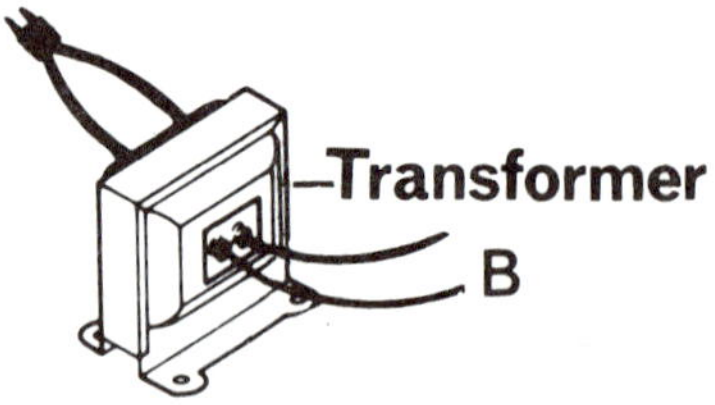

181

If you don't want to use a transformer, use battery Illus. 182. As Illus. 183 indicates, a door bell, buzzer or chime requires running one wire from power source, transformer or battery, to bell, the other line from source to button. Another wire is run from bell to button.

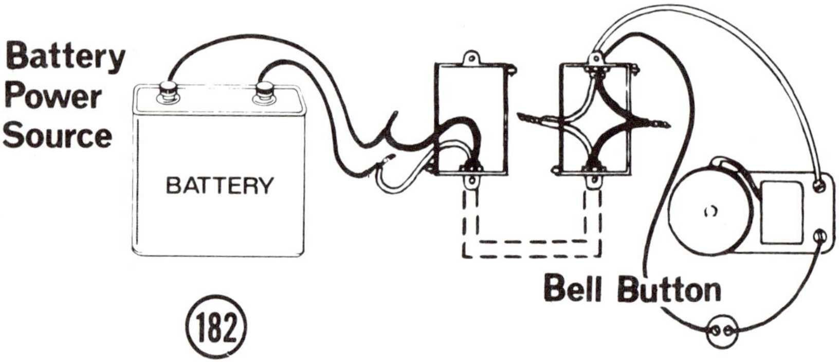

182

If you want to add a buzzer, run hot line from transformer to button, from button to buzzer, from buzzer to other terminal on transformer.

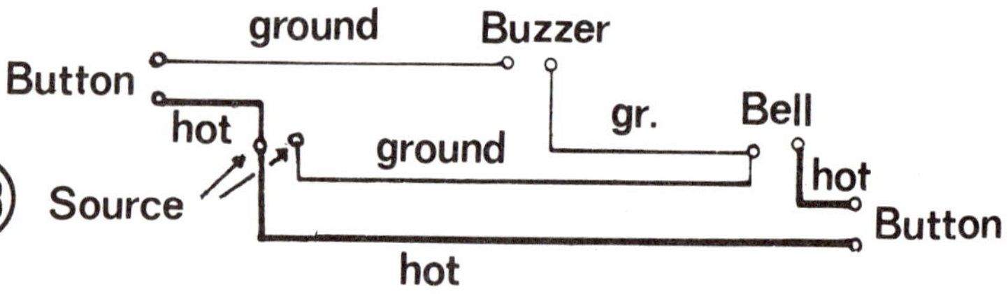

183

CEILING FIXTURE AND LAMP REPAIRS

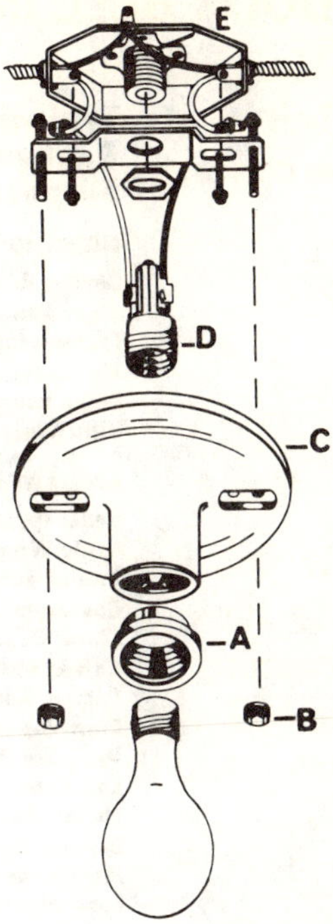

Note name of parts and always purchase exact replacement. To replace interior, remove fuse serving circuit. Remove bulb. Unscrew ring A, and nuts B. Remove base C. Disconnect wires from socket D. Replace with interior. Reassemble lamp holder. Some overhead ceiling lights have an extended stem fastened to E that holds globe in position.

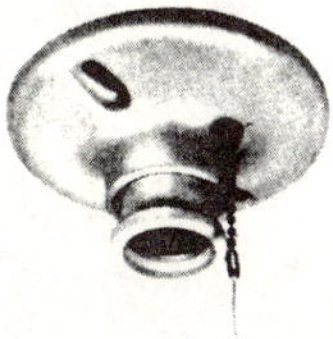

PULL CHAIN CEILING LAMPHOLDER

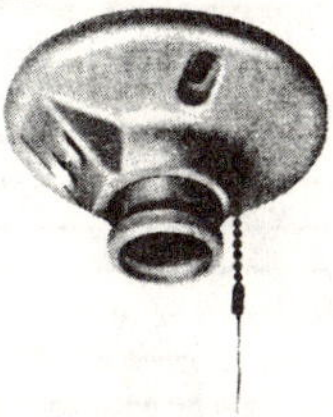

PULL CHAIN CEILING LAMPHOLDER WITH SIDE OUTLET

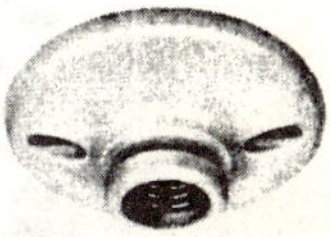

KEYLESS CEILING LAMPHOLDER

PULL CHAIN INTERIOR

PULL CHAIN AND SIDE OUTLET INTERIOR

3-WAY SOCKET

PULL CHAIN SOCKET

PUSH THROUGH SOCKET

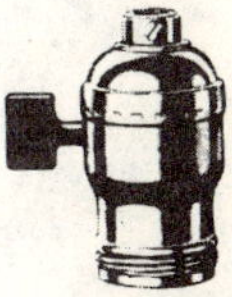

KEY SOCKET

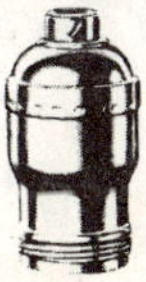

KEYLESS SOCKET

PUSH-THROUGH INTERIOR

KEY INTERIOR

Typical Wattages of Some Lights and Appliances Normally Connected to General Purpose or Plug-in Appliance Circuits

LIGHTING

Ceiling or Wall (each bulb)	40-150 watts
Floor Lamps (each)	150-300 "
Fluorescent Lights (each tube)	15-40 "
Pin-to-Wall Lamps	50-150 "
Table Lamps (each)	50-150 "
Ultra Violet Lamp	385 "

APPLIANCES

Baker (portable)	800-1000 "
Bottle Warmer	95 "
Broiler-Rotisserie	1320-1650 "
Casserole	1350 "
Clock	2 "
Coffee Maker or Percolator	440-1000 "
Coffee Grinder	150 "
Corn Popper	1350 "
Deep Fat Fryer	1350 "
Egg Cooker	500 "
Electric Bed Cover	200 "
Electric Fan (portable)	100 "
Electric Roaster	1650 "
Food Blender	230-250 "
Hair Dryer	235 "
Hand Iron (steam or dry)	1000 "
Heating Pad	60 "
Heated Tray	500 "
Ice Cream Freezer	115 "
Ironer	1650 "
Knife Sharpener	103 "
Lawn Mower	250 "
Mixer	100 "
Portable Heater	1000 "
Radio (each)	100 "
Record-Changer	75 "
Refrigerator*	150 "
Sandwich Grill	660-800 "
Saucepan	1000 "
Sewing Machine	75 "
Shaver	12 "
Skillet	1100 "
Television	300 "
Toaster (modern automatic	up to 1150 "
Vacuum Cleaner	125 "
Ventilating Fan (built-in)	140 "
Waffle Iron	up to 1100 "
Warmer (Rolls, etc.)	100 "
Waxer-Polisher	350 "

*Each time the refrigerator starts it takes several times this wattage for an instant.

*This estimate is based on 120 volt service.

Please Note:

As this book is based on the National Electrical Code (USA), the following modifications must be considered for use in Canada, in order to ensure compliance with the Canadian Electrical Code:

Page 8 — 3rd Paragraph	A standard parallel blade switch.
9 — 1st Paragraph	Safety standards for electrical installations in Canada are contained in the 11th Edition of the Canadian Electrical Code.
— 6th Paragraph	receptacle (Illustration 13)
11 — 5th Paragraph	may only be protected by a 15-
	All circuits require a separate grounding conductor, which may be either bare conductor or preferably green insulated.
Page 15 — 1st Paragraph	ampere fuse or circuit breaker.
Illus. 5	Use Canadian approved panels only.
16 — 4th Paragraph	
Page 18 — 1st Paragraph	The Canadian Electrical Code now requires non-interchangeable plug fuses which may be of the Type S variety as shown in Illustrations 7 and 8, or of the Type C variety. The latter provides non-interchangeability by means of different size tips and the use of rejectors with different size openings.
Page 19 — 2nd Paragraph	Some Canadian panels also have fuse blocks containing 2 plug fuses, and the block must be withdrawn before the fuses are accessible.
Page 21 — Illus. 16	The ground clips illustrated are not acceptable in Canada.
Illus. 17	Three wire to two wire adapters are not recognized by the Canadian Electrical Code.
Page 28 — 1st Paragraph	
2nd Paragraph	Should specify NMD-7 and NMW-9 and NMW-10 instead of NM and NMC.
Illus. 28	Conduit in picture is called electrical metallic tubing or EMT in Canada.
Page 39	Garage Lighting
	Use NMD-7 for dry locations, NMW-9 or NMW-10 for damp locations and NMW-10 for direct burial.
Page 47 — Last paragraph	These ground clips are not acceptable in Canada.

Continued

Page 50 — 2nd Paragraph	The grounding conductor of non-metallic sheathed cable (NM or NMW) must be terminated on the grounding screw of the box. Wrapping it around the cable is not acceptable.
Page 53 — Illus. 86 Illus. 87	Ranges and dryers must be connected by means of receptacles and power supply cords. The cords and receptacles must be of the four-wire type, as the neutral cannot be used for grounding purposes. The designations for caps and receptacles are as follows: Range — 50 ampere, 120/240 volt, 4-wire, 3-pole configuration 14-50R for receptacle and 14-50P for plug cap. Dryer — 30 ampere, 120/240 volt, 4-wire, 3-pole configuration 14-30R for receptacle and 14-30 for plug cap.
Page 61 — Illus. 109 Illus. 110	Conductor joints and splices shall be made with approved wire connectors, or shall be made mechanically and electrically secure and then soldered.
Page 62 — 2nd Paragraph	Use should be made of a double throw switch with load connected to the centre contacts so that both utility supply and generator supply cannot be connected to the load at the same time.
Page 93 — 3rd Paragraph	ULI is not generally recognized in Canada for approval of electrical equipment. This should refer to CSA which is recognized in all ten provinces.
General	Buy and use only those components and fittings that have been certified by CSA and that bear the CSA seal of approval.